Social Computing, Behavioral Modeling, and Prediction

T0137787

Social Computing, Behavioral Modeling, and Prediction

Edited by

Huan Liu

John J. Salerno

Michael J. Young

 Springer

Editors:

Huan Liu
Ira A. Fulton School of Engineering
Arizona State University
Brickyard Suite 501, 699 South Mill Avenue
Box 878809
Tempe, AZ85287-8809, U.S.A.
huanliu@asu.edu

John J. Salerno

Michael J. Young

ISBN-13: 978-1-4419-4597-6   e-ISBN-13: 978-0-387-77672-9

Printed on acid-free paper.

© 2010 Springer Science+Business Media LLC

9 8 7 6 5 4 3 2 1

springer.com

Preface

Social computing is an area of computer science at the intersection of computational systems and social behavior. Social computing facilitates behavioral modeling, which provides a mechanism for the reproduction of social behaviors and subsequent experimentation in various activities and environments. Contemporary development of novel theories and mechanisms in the understanding of social interactions requires cutting edge research borne from interdisciplinary collaboration. This communication is intended as a scientific literary publication meant to embrace a multidisciplinary audience while maintaining scientific depth.

This was the first international workshop on *Social Computing, Behavioral Modeling and Prediction*, held April 1-2, 2008. It provided an interdisciplinary venue that set the stage for sociologists, behavioral scientists, computer scientists, psychologists, anthropologists, information systems scientists, and operations research scientists to exchange ideas, learn new concepts, and develop new methodologies. Activities included seminars from guest speakers and keynote speakers, and oral and poster presentations. The workshop sessions were arranged into four categories: 1) User's views and requirements, 2) The computer scientist's view, 3) The social scientist's view, and 4) Government plans and research initiatives. The overarching goal was to provide a forum to encourage collaborative efforts in order to increase our understanding of social computation and evaluation. The program committee included researchers from Asia, Europe and the USA, and submissions to the workshop came from Norway, China, Germany, Japan, Italy and the USA, spanning industry, academia and national laboratories.

In one key-note address, Sun-Ki Chai represented University of Hawaii's Department of Sociology in, *"Rational Choice Theory: A Forum for Exchange of Ideas between the Hard and Social Sciences in Predictive Behavioral Modeling"*. Scott Atran represented the Institute for Social Research (University of Michigan) in his key-note speech, *"Comparative Anatomy and Evolution of Terrorist Networks"*. Lastly, *"The SOMA Terror Organization Portal (STOP): Social Network and Analytic Tools for Real-Time Analysis of Terror Groups"* was provided by V.S. Subrahmanian (University of Maryland Institute for Advanced Computer Studies). In addition to keynote addresses, invited speaker Michael Hechter (Arizona State University School of Global Studies) addressed "*The Sociology of Alien Rule*".

In addition to keynote and invited speakers, many scholarly ideas were presented at this workshop, and are conveyed in the following papers. This book contains several general themes, which include 1) data analysis on behavioral patterns, 2) modeling cognitive processes, 3) new tools for analyzing social networks, and 4) modeling organizations and structures. These themes will now be expanded upon in the context of the proceedings of this workshop.

Data analysis on behavioral patterns. Insights into the identification and classification of behavior patterns or attributes of individuals within a social network were presented by several research groups. "*Modeling Malaysian public opinion by Mining the Malaysian Blogosphere*", by Brian Ulicny, delineates indicative behavior patterns of prominent individuals of the Malaysian Blogosphere. "*Behavioral Entropy of a Cellular Phone User*", by Dantu et al. demonstrates the ability to capture an individual user's calling behavior based on various patterns, taking randomness levels into account. "*Inferring social network structure using mobile phone data*", by Eagle et al. offers a new method for measurements of human behavior based on proximity and communication data, which allows for prediction of individual-level outcomes such as job satisfaction.

Analyses to define attributes of a group or groups of individuals were also revealed. "*Using topic analysis to compute identity group attributes*", by Booker and Strong, demonstrates a modeling framework that investigates inter-group dynamics. "*Behavior profiling for computer security applications*", by Robinson et al. suggests a new methodology to categorize World Wide Web users based on web browsing activities. The authors have built a hierarchical structure representing key interest areas, which can be collapsed or expanded to include multiple levels of abstraction. Determinable behavior norms can be analyzed and user fingerprints can be created. Possible security applications include detection of policy violations, hacker activity and insider threat.

Behavior is also influenced by culture, which contributes greatly to group dynamics. "*Computational models of multi-national organizations*", by Levis et al. illustrates an algorithm that takes cultural contributions into account. On a similar note, "*Stochastic Opponent Modeling Agents: a case study with Hezbollah*", by Mannes et al. demonstrates how SOMA can be utilized to reason about cultural groups - in this case, Hezbollah. SOMA derived rules for Hezbollah and described key findings.

Cognitive processes of learning, perception, reasoning, intent, influence and strategy. A great deal of research is being performed on computation and modeling of human cognitive processes. "*Particle swarm social model for group social learning in adaptive environment*", by Cui et al. demonstrates modeling of social learning for self-organized groups in an adaptive environment. This research provides insight into understanding the dynamics of groups such as on-line communities or insurgents. "*Conceptualizing trustworthiness mechanisms for countering insider threats*", by Shuyuan Mary Ho, demonstrates that trustworthiness as determined by perception in an on-line community is a significant element in corporate personnel security. *"Metagame strategies of nation-states, with application to cross-strait relations"*, by Chavez and Zhang, investigates models for strategic reasoning in gaming, where players mutually predict the actions of other players. This approach is highly applicable to the analysis of reasoning within nation-states. *"Reading between the lines: human-centered classification of communication patterns and intentions"*, by Stokar von Neuforn and Franke, identifies communication patterns for qualitative text-context analysis, which aims to derive purpose, context and tone in order to infer intention. Lastly, *"Clustering of trajectory data obtained from soccer game records"*, by Hirano and Tsumoto, summarizes the strategy of game players. Results demonstrate patterns that may be associated with successful goals.

As briefly touched upon above, there are many cognitive complexities associated with human behavior. *"Human behavioral modeling using Fuzzy and Dempster-Shafer Theory"*, by Ronald Yager, combines fuzzy sets for the representation of human-centered cognitive concepts with Dempster-Shafer Theory to include randomness in the fuzzy systems. This combined methodology allows for the construction of models that include cognitive complexities and unpredictability in the modeling of human behavior.

Tools for analysis of social networks. By far, the largest general category presented at this workshop included tools and approaches for the analysis of social networks. *"Social network analysis: tasks and tools"*, by Loscalzo and Yu, provides an evaluation of tools developed for social analysis techniques, and how these tools address given tasks in social network analysis. *"Automating frame analysis"*, by Sanfilippo et al, presents an approach to the representation, acquisition, and analysis of frame evidence. This approach provides content analysis, information extraction and semantic search methods in order to automate frame annotation. *"A composable discrete-time cellular automaton formalism"*, by Mayer and Sarjoughian, provides support for extensive use of cellular automata in simulating complex heterogeneous systems in formal model specification.

A multidisciplinary approach to social network analysis was presented in, *"Integrating multi-agent technology with cognitive modeling to develop an insurgency information framework (IFF)"*, by Bronner and Richards. Their work aims to develop a decision-aiding model for behavior strategy analysis of Iraqi insurgents. Specifically, the authors combine sophisticated software engineering technology and cognitive modeling tools to deliver a tool that provides military leaders aide in approaching insurgency problems.

Tools for analysis of inter-group relations were also presented. *"Modeling and supporting common ground in geo-collaboration"*, by Convertino et al., proposes the introduction of process visualizations and annotations tools to support common ground in the sharing and managing of knowledge in geo-collaboration. *"An approach to modeling group behaviors and beliefs in conflict situations"*, by Geddes and Atkinson, describes an approach to modeling group responses to planned operations over extended periods of time in modern military operations.

Finally, research concerned with characterization of sub-network dynamics was presented. *"Where are the slums? New approaches to urban regeneration"*, by Murgante et al., introduces a method for producing a more detailed analysis for urban regeneration policies and programs. This method identifies high priority areas for intervention, increasing efficiency and investment effectiveness in urban regeneration programs. *"Mining for Social Process Signatures in Intelligence Data Streams"*, by Savell and Cybenko, introduces a robust method for defining an active social network, and identifying and tracking the dynamic states of subnetworks.

Modeling organization and structure. *"Community detection in a large real-world social network"*, by Steinhauser and Chawla, demonstrates that a simple thresholding method with edge weights based on node attributes is sufficient in identifying community structure. *"An ant colony optimization approach to expert identifications in social networks"*, by Ahmad and Srivastava, demonstrates an efficient way to identify experts in a network and direct pertinent queries to these experts.

This book conveys the latest advances in social computing, as communicated at our first international workshop on Social Computing, Behavioral Modeling and Prediction. While the topics presented at this workshop were widespread, the methods and analyses described have broad potential for applicability. We dearly thank and are deeply indebted to all the distinguished scientists who participated.

Phoenix, Arizona,   *Dr. John Tangney, ONR*
April, 2008   *Dr. Judith M. Lytle, ONR*

Acknowledgements

We would like to first express our deep gratitude to Mr. John Graniero, Director of AFRL/II for encouraging us to organize this interdisciplinary workshop on emerging topics associated to social computing, behavioral modeling, and prediction, and for providing ideas and funding to get it started. We truly appreciate the timely help and assistance provided by Melissa Fearon and Valerie Schofield of Springer US to make the publication of the proceedings possible in a relatively short time frame. We thank the authors wholeheartedly for contributing a range of research topics showcasing current and interesting activities and pressing issues. Due to the space limit, we could not include as many papers as we wished. We thank the program committee members to help us review and provide constructive comments and suggestions. Their objective reviews greatly improved the overall quality and content of the papers. We would like to thank our keynote and invited speakers by presenting their unique research and views. We thank the members of the organizing committee for helping to run the workshop smoothly; from the call for papers, the website development and update, to proceedings production and registration. We also received kind help from Kathleen Fretwell, Helen Burns, Audrey Avant, and Robert Salinas of School of Computing and Informatics, ASU in tedious but important matters that are transparent to many of us. Last but not least, we sincerely appreciate the efforts of Drs. John Tangney, and Judith M. Lytle to help and guide this workshop and relevant research activities with their insights and visions.

Without the assistance from many, it would be certain that the workshop and proceedings would not be possible. We thank all for their kind help.

Contents

List of Contributors

Ashraf M. AbuSharekh, George Mason University
Muhammad Ahmad, University of Minnesota - Twin Cities
Michele Atkinson, Applied Systems Intelligence, Inc.
Vincent Berk, Dartmouth College
Lashon Booker, The MITRE Corporation
LeeRoy Bronner, Morgan State University
John M. Carroll, The Pennsylvania State University
Giuseppe Las Casas, University of Basilicata
Sun-Ki Chai, University of Hawaii
Alex Chavez, University of Michigan
Nitesh Chawla, University of Notre Dame
Gregorio Convertino, The Pennsylvania State University
Xiaohui Cui, Oak Ridge National Laboratory
George Cybenko, Dartmouth College
Maria Danese, University of Basilicata
Gary Danielson, Pacific Northwest National Laboratory
Ram Dantu, University of North Texas
Nathan Eagle, MIT / Santa Fe Institute
Katrin Franke, Gjøvik University College, Norway
Lyndsey Franklin, Pacific Northwest National Laboratory
Craig H. Ganoe, The Pennsylvania State University
Norman Geddes, Applied Systems Intelligence
Rebecca Goolsby, Office of Naval Research
Michael Hechter, Arizona State University
Shoji Hirano, Shimane University
Shuyuan Mary Ho, Syracuse University
Blaine Hoffman, The Pennsylvania State University
Husain Husna, University of North Texas
Smriti K. Kansal, George Mason University
David Lazer, Harvard University

Alexander H. Levis, George Mason University
Steven Loscalzo, Binghamton University
Judith Lytle, Office of Naval Research
Aaron Mannes, University of Maryland College Park
Gary Mayer, Arizona State University
Liam McGrath, Pacific Northwest National Laboratory
Andrew Mehler, Stony Brook University
Mary Michael, University of Maryland College Park
Nicholas Mileson, Pacific Northwest National Laboratory
Beniamino Murgante, University of Basilicata
Daniela Stokar von Neuforn, Brandenburg University of Applied Sciences,
A. Erkin Olmez, George Mason University
Amy Pate, University of Maryland College Park
Robert Patton, Oak Ridge National Laboratory
Alex (Sandy) Pentland, MIT
Santi Phithakkitnukoon, University of North Texas
Thomas Potok, Oak Ridge National Laboratory
Laura Pullum, Lockheed Martin Corporation
Akeila Richards, Morgan State University
Roderick Riensche, Pacific Northwest National Laboratory
David Robinson, Dartmouth College
Antonio Sanfilippo, Pacific Northwest National Laboratory
Hessam Sarjoughian, Arizona State University
Robert Savell, Dartmouth College
Sajid Shaikh, Kent State University
Steven Skiena, Stony Brook University
Amy Sliva, University of Maryland College Park
Jaideep Srivastava, University of Minnesota - Twin Cities
Karsten Steinhaeuser, University of Notre Dame
Gary Strong, Johns Hopkins University
V.S. Subrahmanian, University of Maryland
Stephen Tratz, Pacific Northwest National Laboratory
Jim Treadwell, Oak Ridge National Laboratory
Shusaku Tsumoto, Shimane University
Brian Ulicny, VIStology, Inc.
Jonathan Wilkenfeld, University of Maryland
Anna Wu, The Pennsylvania State University
Ronald Yager, Iona College
Lei Yu, Binghamton University
Jun Zhang, AFOSR
Xiaolong (Luke) Zhang, The Pennsylvania State University

Organizing Committee

Huan Liu, John Salerno, Michael Young, Co-Chairs
Nitin Agarwal, Proceedings Chair
Lei Yu, Poster Session Chair
Hessam Sarjoughian, Treasurer
Lei Tang and Magdiel Galan, Publicity Chairs

Program Committee

Edo Airoldi, Princeton University
Chitta Baral, ASU
Herb Bell, AFRL
Lashon Booker, MITRE
Sun-Ki Chai, University of Hawaii
Hsinchun Chen, Univiversity of Arizona
David Cheung, University of Hong Kong
Gerald Fensterer, AFRL
Laurie Fenstermacher, AFRL
Harold Hawkins, ONR
Michael Hinman, AFRL
Hillol Kargupta, UMBC
Rebecca Goolsby, ONR
Alexander Levis, GMU
Huan Liu, ASU
Mitja Lustrek, IJS, SI
Terry Lyons, AFOSR

Hiroshi Motoda, AFOSR/AOARD
Dana Nau, University of Maryland
John Salerno, AFRL
Hessam Sarjoughian, ASU
Jaideep Srivastava, UMN
Gary Strong, Johns Hopkins University
V.S. Subrahmanian, University of Maryland
John Tangney, ONR
Tom Taylor, ASU
Ray Trechter, Sandia National Laboratory
Belle Tseng, NEC Labs
Ronald Yager, Iona College
Michael Young, AFRL
Philip Yu, IBM T.J. Watson
Jun Zhang, AFOSR
Jianping Zhang, MITRE
Daniel Zeng, University of Arizona

Rational Choice Theory: A Forum for Exchange of Ideas between the Hard and Social Sciences in Predictive Behavioral Modeling

Sun-Ki Chai[†]

[†] sunki@hawaii.edu, Department of Sociology, University of Hawai`i, Honolulu HI

Abstract The rational choice model of human behavior provides general assumptions that underlie most of the predictive behavioral modeling done in the social sciences. Surprisingly, despite its origins in the work of eminent mathematicians and computer scientists and its current prominence in the social sciences, there has been relatively little interest among hard scientists in incorporating rational choice assumptions into their agent-based analysis of behavior. It is argued here that doing so will introduce greater theoretical generality into agent-based models, and will also provide hard scientists an opportunity to contribute solutions to known weaknesses in the conventional version of rational choice. This in turn will invigorate the dialogue between social scientists and hard scientists studying social phenomena.

1 Introduction

It is very difficult to survey recent work within the social sciences on social computing, behavioral modeling, and prediction in any reasonably concise fashion. For many decades, mathematical, predictive models of behavior have characterized almost all of economic theory and major parts of other social science disciplines as well, particularly within the broad framework of the theoretical approach known as rational choice theory. So in a sense, formal work on predictive behavioral modeling makes up a large portion of the social sciences as a whole.

Rather than attempting to cover such an expanse of research, I will limit myself to investigating the base set of assumptions about human nature that underlies much if not most of this social science literature, the assumptions of rational choice. The next section of this paper will discuss the origins of the rational choice model during an era in which the social sciences were heavily influenced by developments in the "hard" sciences. Following this, I will discuss debates on the strengths and weaknesses of the rational choice model, and attempts to improve it. In the concluding section, I will argue that this debate over rational choice presents an opportunity for hard scientists studying human behavior to improve the generality of their own models while once again contributing to the development of the social sciences.

2 The Origins of the Rational Choice Model in Mathematics and Computer Science

The use of hard science methods as a basis for the prediction of behavior in the social sciences is nothing new. For the leaders of the "marginalist" revolution of the late 19th century that led the modern form of mathematical economics, physics provided a paradigm to be emulated (Mirosky, 1991). While late 19th century sociologists were more circumspect about adopting scientific over humanistic methods, as typified by Weber's concept of *verstehen*, the general desire to build "science of society" was widely accepted (Lepenies and Hollingdale, 1988).

More concretely, and of greater direct relevance to the issues of this conference, foundational work in expected utility theory and game theory came about as a collaboration between the computer scientist/mathematician John von Neumann and the economist Oskar Morgenstern (1944). Their basic formal frameworks, augmented by the later work of Leonard Savage, John Nash (both mathematicians), and others, have provided the technical apparatus, as well as some of the basic assumptions, for most mathematical predictive theories of behavior in the social sciences, which in turn go under the general rubric of formal rational choice theory.

However, despite the clear heritage of formal rational choice modeling in mathematics and computer science, subsequent development of the rational choice literature has not continued to join the social and hard sciences, but more often to separate the two. The bulk of subsequent development in the rational choice approach has taken place in the field of economics, with contributions from political science, sociology, and occasionally other related social sciences. Dialogue and collaboration with the hard sciences has been much less common than one might expect given the origins of the theoretical approach.

One major exception is evolutionary biology, where the pioneering work of John Maynard Smith (1982) triggered a fruitful dialog with economists that helped turned evolutionary game theory (Weibull, 1997) into a truly interdisciplinary field of technical research, and generated within economics such literatures as the evolutionary theory of the firm (Nelson and Winters, 2006 [1988]) and the theory of learning in organizations (Fudenberg and Levine, 1998).

Another exception is operations research, where game theory has long stood beside linear and linear programming, queueing theory, etc. as one of the basic pallete of analytical techniques in the field (Parthasarathy et al, 1997). However, much of the interest in game theory within operations research has been in cooperative rather than non-cooperative game theory, which has somewhat reduced its impact on the social sciences. Moreover, it has been argued that while game theory has penetrated the conversation of OR applied scholars, less emphasis has been placed on producing foundational theoretical research (Shubik, 2002).

A specific example of early cross-fertilization is Robert Axelrod's pioneering use of computer simulations to investigate the relative fitness of decision strategies in iterated Prisoner's Dilemma games (Axelrod, 2005 [1984]; 1997). More recently, there has been a burgeoning in the importation of ideas from the hard sciences, particularly computer science, to enrich social science mathematical methods. The use of agent-based social simulations and other techniques arising from artificial intelligence/cognitive science has become widely accepted in mainstream sociology (Bainbridge et al, 1994; Macy and Willer, 2002) and is making inroads into economics, particularly under rubric of behavioral economics (Camerer et al, 2003), which not incidentally takes Herbert Simon as one of its founding fathers.

3 Debate over Rational Choice within the Social Sciences

To some extent, this renewal of interest in models from the hard sciences is a result of frustration with the weaknesses of the conventional version of the rational choice model, whose assumptions about human behavior have dominated formal behavioral modeling in the social sciences. Its base assumptions can be summarized informally as the idea that an actor seeks to maximize preferences, usually reflected by utility function, a quantity that reflects the desirableness of conceivable outcomes, and does so in light of certain beliefs that she/he holds, which are usually represented as subjective probabilities, sometimes contingent, about the state of the environment. The existence of a utility function reflects certain inherent assumptions about the properties of preferences, most notably that they constitute a strict order having the properties of completeness/connectedness, irreflexivity, acyclicity, and transitivity. Beliefs are constrained to be consistent with each other and the usual axioms of probability theory (Chai, 1997; 2001, chap. 1).

In practice, these base assumptions are not enough to generate predictions about behavior, and must be augmented with auxiliary assumptions or methodologies that determine the substance of preferences and beliefs. In the conventional version of rational choice, preferences are assumed to be unidimensional, self-regarding, materialistic, and isomorphic across actors. In other words, preferences for all individuals can be represented by their own endowment of a single material good, typically money. Other typical additions to the base assumptions involve the utility attached to uncertain outcomes (expected utility) and the way that utility is cumulated over time (exponential discounting of future utility).

Also in the conventional version, beliefs are based on what can be called an "information assumption", i.e. that beliefs derive solely from facts that an actor can observe directly, as well as propositions that can be inferred from such facts through logic or probability theory. The information assumption is a formal device that allows one to infer the totality of an actors beliefs from their past and present position (defined both spatially and socially) in the environment. One other addition to belief assumptions is common knowledge of rationality, the idea that actors are all aware of each other's rationality (in the sense being described here), a crucial assumption for generating predictions from the game theory formalism.

The great strength of conventional rational choice can be summarized by its combination of generality, decisiveness, parsimony, and relative accuracy. Generality refers to the applicability of the conventional rational choice model to virtually any environment and choice situation. Because its assumptions encompass the necessary characteristics for representing choice using game theory and expected utility formalizations, it can be applied to choice situation that can be represented using these formalizations.

In most such environments, conventional rational choice assumptions not only can be applied, but will generate predictions that specify a set of possible outcomes that is significantly more constrained than the set of conceivable outcomes. This is the quality of decisiveness, and it ensures that across these environments, predictions will not only tend to be falsifiable (as long as the outcomes represented are operationalizable), but also non-trivial, ruling out outcomes that are not *ipso facto* impossible or improbable.

The conventional assumptions, while more complex than the base ones, are parsimonious in that only a small number of variables that have to be considered in generating predictions about behavior. Indeed, actor preferences and beliefs are both endogenous to (specified from within) the conventional model, hence the only information that is needed to generate behavioral predictions within any environment are the past and present positions of the actor within the environment.

Finally, the predictions that conventional rational choice model has generated, despite the seeming handicap of carrying a single set of assumptions across the full domain of choice environments, has often generated predictions that are often both provocative and confirmed by empirical evidence. While most of these predictions have been its "home" environment of economic behavior, recent efforts have greatly expanded the realm of prediction to areas far afield. The recent phenomenon of "freakonomics" (Leavitt and Dubner, 2006) is simply the most recent manifestation of a long process of expansion in the domain of rational choice, a process that has been referred to most commonly, by both proponents and opponents, as "economic imperialism" (Radnitzky and Bernholz, 1987; Lazear, 2000).

No other existing major theoretical approach equals conventional rational choice in meeting this combination of criteria, and this has been seen as justifying the approach's preeminence in social science predictive behavioral modeling. Put together, the criteria ensure that findings arising from the application of the conventional rational choice model cumulate into law-like statements based on an internally consistent view of human nature and behavior. Because of this cumulability, it is relatively easy to generalize from rational choice findings to formulate new, testable hypotheses in arenas where the model has yet to be applied, an important factor when the model is the basis for predictive technology that will may be used under future, unforeseen circumstances. Furthermore, because these hypotheses arise from the application of a general model that has been confirmed under a multiplicity of environments, there is greater reason to believe that the hypotheses will be accurate than if they were simply formulated in an *ad hoc* fashion. As such, it is without contest the preeminent embodiment in the social sciences of the nomothetic method associated with the natural sciences (Windelband, 1894).

Despite these positive qualities, the conventional rational choice approach has over the past decade been subject to a host of criticisms that have exposed its major weaknesses in the harshest light. While it has always been accepted that the conventional assumptions comprise an extreme simplification of human psychology, the true extent to which this hampers its ability to generate accurate predictions has become increasingly clear.

The approach been known to generate major predictive anomalies across a wide set of environments. These include anomalies of indecisiveness, where theories based on conventional rational choice cannot generate predictions that substantially reduce the set of all feasible behavioral outcomes because the actors involved do not have sufficient information to decide from logical alone which choices are better than others. This has been found most frequently in environments where actors are engaged in repeated interactions of uncertain duration (Fudenberg and Maskin, 1986) or where there is a need for actors to coordinate behaviors with one another. They also include anomalies of inaccuracy, where conventional theories predict outcomes that differ systematically from observed behavior. This is found most often in "social dilemma" conditions where cooperation is mutually beneficial but cannot be enforced by individual reward (Henrich et al, 2004; Gintis et al, 2006). These anomalies are seen as being linked to underlying shortcomings with the motivational and cognitive assumptions of the conventional approach.

There have been two basic policies put forward to addressing these shortcomings. The first has been to advocate abandoning the rational choice approach altogether. The second has been to try keeping the baby while throwing out the bathwater, attempting to retain the strengths of the conventional approach while improving its predictive ability by augmenting it with new assumptions and techniques. In particular, the focus for the latter has been on replacing or augmenting its assumptions. This in turn can be done by allowing for greater variations in the types of beliefs and values that actors can hold, as well as allowing for change over time in these attributes of actors, often reflecting larger changes in culture. However, this must be done in a general, rather than *ad hoc,* manner in order to retain the strengths of the model.

In effect, any attempt to do so will constitute an attempt to insert culture into rational choice. This is not a simple task, and in the past it was taken for granted that the cultural and rational choice analysis were inherently incompatible (Harsanyi, 1969; Barry, 1970). While this is no longer accepted to be the case, and the way in culture can be inserted into rational choice are quite varied, it is much more difficult to do so in a way that retains the generality and decisiveness of the conventional approach.

There are arguably two primary ways of incorporating culture in such a general and decisive fashion. The first is the creation of general cultural dimensions and typologies that limit the kinds of cultural configurations that exist within a society. The second is an general endogenous theory of culture formation that can predict variations in preferences and beliefs across individuals and over time.

An initial attempt to combine both types of methods is the grid-group/coherence approach. One component makes use of the grid-group cultural dimensions developed by the eminent social anthropologist Mary Douglas (1970; 1984), adapting them to make them compatible with the base assumptions of rational choice analysis of behavior (Chai and Wildavsky 1994; Chai and Swedlow 1998). The other component introduces a coherence model of preference and belief formation that is based on the synthesis and formalization of various prominent theories of self found in psychology, sociology, and anthropology (Chai, 2001). However, this approach certainly is not the only possible way of incorporating cultural in a general fashion, and other efforts are awaited.

4 An Opportunity for Synergy in Reformulating Rational Choice

As the papers presented at this conference attest, there is no doubt that behavioral modeling has enjoyed a productive boom among scholars in the hard sciences, particularly in the fields of computer science, electrical and industrial engineering, and biology. This work has brought a new level of technical rigor to the formal analysis of behavior, as well as development of an array of technical tools for analyzing and predicting behavior.

Perhaps not surprisingly, the dominant mode of analysis for computer scientists, engineers, mathematicians, and biologists engaging in prediction of human behavior has been computational agent-based modeling. Somewhat less expectedly, hard scientists have tended put little effort into coming to some agreement about the basic general assumptions regarding human behavior that will underlie their use of these tools. Models of behavior tend to be designed to suit particular environments being analyzed, and there is little debate over the proper core assumptions that should be held in common by such models. Agents are assigned decision rules and strategies without much of an attempt to link those rules to an explicit, much less unified, view of human nature. This is surprising in light of the fact that the hard sciences, much more than the social sciences, have traditionally been viewed as being governed by nomothetic approach to knowledge (Windelband, 1894), i.e. one that is aimed at creating general laws of nature that apply across environments and circumstances.

It is here that great potential exists for fruitful dialogue between hard and social scientists. The base assumptions of the rational choice model provide by far the most widely tested platform on which to build a general model of human behavior, and can easily be adapted to many if not all of the agent-based modeling techniques and formalizations that have been developed over the last several years.

However, the ongoing crisis of conventional rational choice provides an opportunity for hard scientists to not only to borrow the social scientist's preeminent general model, but also to help improve it by using their technical sophistication to suggest formal solutions to the issue of culture that will fit the criteria of generality and decisiveness, yet also be able to account for patterns found in behavioral data across environmental domains. In a way, this will return us full circle to the days of von Neumann and Morgenstern, when the boundaries between the hard and social science intellectual commities seemed more permeable than they have since.

References

1. Axelrod, Robert (2006 [1984]) The Evolution of Cooperation. Perseus Books, New York.
2. Axelrod, Robert (1997) The Complexity of Cooperation: Agent-Based Models of Competition and Collaboration. Princeton University Press, Princeton NJ.
3. Bainbridge, William Sims, Edward E. Brent, Kathleen M. Carley, David R. Heise, Michael W. Macy, Barry Markovsky, and John Skvoretz (1994) Artificial Social Intelligence. Annual Review of Sociology 20: 407-436.
4. Barry, Brian (1970) Sociologists, Economists and Democracy. Collier-MacMillan, London.
5. Camerer, Colin F., George Loewenstein, and Matthew Rabin, ed. (2003) Advances in Behavioral Economics. Princeton University Press, Princeton NJ.
6. Chai, Sun-Ki (1997) Rational Choice and Culture: Clashing Perspectives or Complementary Modes of Analysis? In: Richard Ellis and Michael Thompson, ed., Culture Matters 45-56. Westview Press, Boulder, CO.
7. Chai, Sun-Ki (2001) Choosing an Identity: A General Model of Preference and Belief Formation. University of Michigan Press, Ann Arbor, MI.
8. Chai, Sun-Ki and Brendon Swedlow, Ed. (1998) Culture and Social Theory. Transaction Publishers, New Brunswick NJ.

9. Chai, Sun-Ki and Aaron Wildavsky (1994) Cultural Theory, Rationality and Political Violence. In: Dennis J. Coyle and Richard J. Ellis, Eds. Politics, Culture and Policy: Applications of Cultural Theory. Westview Press, Boulder.

10. Douglas, Mary (1970) Natural Symbols: Explorations in Cosmology. Pantheon, New York.

11. Douglas, Mary (1982) In the Active Voice. Routledge and Keegan Paul, London.

12. Fudenberg, Drew, and Eric Maskin (1986) The Folk Theorem in Repeated Games with Discounting or with Incomplete Information. Econometrica 54:533-54.

13. Fudenberg, Drew and David K. Levine (1998) The Theory of Learning in Games. MIT Press, Cambridge, MA.

14. Gintis, Herbert, Samuel Bowles, Robert T. Boyd, and Ernst Fehr (2006) Moral Sentiments and Material Interests: The Foundations of Cooperation in Economic Life. MIT Press, Cambridge MA.

15. Harsanyi, John (1969) Rational Choice Models of Political Behavior vs. Functionalist and Conformist Theories. World Politics 21 (June):513-38.

16. Henrich, Joseph, Robert Boyd, Samuel Bowles, Colin Camerer, Ernst Fehr, and Herbert Gintis (2004) Foundations of Human Sociality: Economic Experiments and Ethnographic Evidence from Fifteen Small-Scale Societies. Oxford University Press, New York.

17. Lazear, Edward P. (2000) Economic Imperialism. Quarterly Journal of Economics 115,1 (Feb):99-146.

18. Levitt, Steven D. and Stephen J. Dubner (2006) Freakonomics: A Rogue Economist Explores the Hidden Side of Everything. William Morrow, New York.

19. Lepenies, Wolf and R. J. Hollingdale (1988) Between Literature and Science: The Rise of Sociology. Cambridge University Press, Cambridge.

20. Macy, Michael W. and Robert Willer (2002). From Factors To Actors: Computational Sociology and Agent-Based Modeling. Annual Review of Sociology 28, 143-66.

21. Maynard Smith, John (1982) Evolution and the Theory of Games. Cambridge University Press, Cambridge.

22. Mirowski, Philip (1991) More Heat than Light: Economics as Social Physics, Physics as Nature's Economics. Cambridge University Press, Cambridge.

23. Nelson, Richard R. and Sidney G. Winter (2006) An Evolutionary Theory of Economic Change. Belknap Press, Cambridge, MA.

24. Parthasarathy, T., B. Dutta, J.A.M. Potters, T.E.S. Raghaven, D. Ray, A. Sen (1997) Game Theoretical Applications to Economics and Operations Research. Springer, Berlin.

25. Radnitzky, Gerard, and Peter Bernholz, eds. (1987) Economic Imperialism: The Economic Approach Applied Outside the Field of Economics. Paragon House, New York.

26. Shubik, Martin (2002) Game Theory and Operations Research: Some Musings 50 Years Later. Operations Research 50,1 (Jan/Feb):192-196.

27. Thompson, Michael, Richard Ellis and Aaron Wildavsky (1990) Cultural Theory. Westview Press, Boulder CO.

28. von Neumann, John, and Oskar Morgenstern (1944) Theory of Games and Economic Behavior. Princeton University Press, Princeton NJ.

29. Weibull, Jörgen W. (1997) Evolutionary Game Theory. MIT Press, Cambridge, MA.

30. Windelband, Wilhelm (1894) Geschichte und Naturwissenschaft, Heitz, Strassburg

The SOMA Terror Organization Portal (STOP): social network and analytic tools for the real-time analysis of terror groups

Amy Sliva[†], V.S. Subrahmanian[†], Vanina Martinez[†], and Gerardo I. Simari[†]

[†]{asliva,vs,mvm,gisimari,asliva}@cs.umd.edu, Lab for Computational Cultural Dynamics, Institute for Advanced Computer Studies, University of Maryland, College Park, MD 20742, USA

Abstract Stochastic Opponent Modeling Agents (SOMA) have been proposed as a paradigm for reasoning about cultural groups, terror groups, and other socio-economic-political-military organizations worldwide. In this paper, we describe the SOMA Terror Organization Portal (STOP). STOP provides a single point of contact through which analysts may access data about terror groups world wide. In order to analyze this data, SOMA provides three major components: the SOMA Extraction Engine (SEE), the SOMA Adversarial Forecast Engine (SAFE), and the SOMA Analyst NEtwork (SANE) that allows analysts to find other analysts doing similar work, share findings with them, and let consensus findings emerge. This paper describes the STOP framework.

1 Introduction

Stochastic Opponent Modeling Agents introduced in [1,2,3] were introduced as a paradigm for reasoning about any group G in the world, irrespective of whether the group is a terror group, a social organization, a political party, a religious group, a militia, or an economic organization. SOMA-rules have been used to encode the behavior of players in the Afghan drug economy [4] as well as various tribes along the Pakistan-Afghanistan border [5], as well as terror groups such as Hezbollah [6]. Continuing studies are expected to track over 300 terror groups by end 2008.

The study of terror groups in the national security community is hampered by many major problems.

1. **Lack of timely, accurate data.** Data about the terror groups is collected manually. This causes any data to be either incomplete or out of date (usually both). One of the best known collections of data about terror groups is the "Minorities at Risk Organizational Behavior" (MAROB) data set [7,8] which tracks approximately 400 properties of ethnopolitical

9

groups representing 284 ethnic groups worldwide. As mentioned in [5], such manual data collections are incomplete (because the manual coders of this data are only able to process certain numbers of articles in a limited number of languages) and coarse grained (because the variables are classified into categories rather than specific numbers – for instance, the number of deaths caused by a group might be classified as "none", "low", "medium" or "high" instead of giving a reasonably accurate estimate). Moreover, they are usually out of date (Minorities at Risk coding is currently complete only until 2003), leaving the data to be "non-actionable".

2. **Lack of behavioral models.** Analysts today are forced to make up behavioral models in a slow and painstaking manner, The SOMA Extraction engine (SEE), a portion of STOP, extracts behavioral rules automatically from the MAROB data. To date, it has extracted rules on approximately 23 groups with another 2 or 3 groups being analyzed and completed each week.

3. **Lack of a social network.** Analysts are often unaware of what other analysts have found useful. There is no computational mechanism we have seen to date that allows analysts to share their experiences and forecasts for a particular group. In fact, analysts are often in the dark about who else is looking at a certain group – either from the same, or different viewpoint as themselves. The **SOMA Analyst NEtwork (SANE)** provides a social networking framework within which analysts can browse data about the groups of interest, rate the accuracy of the data, browse rules extracted by SEE, rate the rules extracted by SEE, and see the predictions produced by SAFE.

4. **Lack of a forecast engine.** The SOMA Adversarial Forecast Engine (SAFE) uses a rich probabilistic foundation with no unwarranted independence assumptions in order to forecast the most probable sets of actions that a terror group might take in a given situation. Using SAFE, analysts can analyze the current situation and/or hypothesize new situations based on possible actions the US might be contemplating.

We will not be discussing problem (1) above in much detail in this paper because our CARA architecture described in [9] already addresses many of the problems of dealing with real time data. Components of CARA such as OASYS [10] and T-REX [11] extract valuable data from open sources in real time.

2 SOMA Extraction Engine

The SOMA Extraction Engine (SEE) is intended to derive SOMA-rules automatically from real-time sources such as T-REX [10] and OASYS[11]. These systems automatically extract several kinds of events. They are being tuned to extract social science codes such as those used in the MAROB effort. Till that is complete, the SOMA extraction engine is directly working with the MAROB data to extract rules.

A SOMA rule about a group G has the form

$$<Action>:[L,U] \text{ if } <Env\text{-}Condition>$$

Where:

- <Action> is an action that the group took (such as KIDNAP)
- <Env-Condition> is a logical conjunction of elementary conditions on the environmental attributes. An elementary condition associated with the environmental attribute A is an expression of the form A op value where op is in the set { =, <=, >= }.
- [L,U] is a closed sub-interval of the [0,1] interval.

The above rule says that in any year when the <Env-Condition> is true, there is a probability between L and U that the group took the action stated in the rule. The rule below is an example of a rule that we extracted about Hezbollah.

KIDNAP: [0.51,0.55] if solicits-external-support & does not advocate democracy.

This rule says that in years when Hezbollah both solicited external support and did not promote democratic institutions, there was a 51 to 55% probability that they engaged in kidnapping as a strategy.

The SEE Algorithm works in accordance with the sketch given below. It takes as input, a group G, along with a table $Tab(G)$, for the group. This table contains one column for each attribute of this group, and one row for each year. The attributes are split into *action* attributes (which describe the actions the group took in a given year) and *environmental* attributes (which describe the environment in which the group functioned during that year – these can include actions that *other* organizations took against group G). SEE takes an action name a and a code c for that action as input (e.g. KIDNAP may be an action and the code 0 might indicate the code for no kidnappings). Based on these inputs, SEE attempts to find all conditions *on the environmental attributes* that predict KIDNAP=0 (i.e. that kidnappings will not be resorted to by the group) with a high degree of probability. Clearly, we are only interested in conditions which, when true, predict KIDNAP=0 with high probability, and which when false, predict KIDNAP=1 with high probability. In conditional probability terms, we are interested in finding conditions C on the environmental attributes such that abs(P(KIDNAP=0|C) − P(KIDNAP=0|~C)) is high. As usual, P(KIDNAP=0|C) denotes the conditional probability of the value of the KIDNAP attribute being 0 in a given year, given that condition C holds for that year.

Of course, even when we find such a condition C above, we would like to ensure adequate *support*, i.e. to ensure that there is a sufficiently large number of years in which both C and ~C are true.

The SEE system implements the algorithm in [2] to ensure that the conditions found satisfy both these requirements.

As of the time of writing this paper (early Dec. 2007), SEE has extracted behavioral models of 23 groups worldwide. The list includes 8 Kurdish groups spanning Iran, Turkey, and Iraq (including groups like the PKK and KDPI), 8 Lebanese groups (including Hezbollah), several groups in Afghanistan, as well as several other Middle Eastern groups.

3 SOMA Analyst NEtwork (SANE)

STOP requires an analyst to register and log in prior to accessing the system. Once the user is logged in, she proceeds via four phases.

Phase 1. Selecting a Group. In this phase, the user selects a group to study by examining a drop down list of groups that have been processed so far. Figure 1 below shows this screen when the user has selected the PKK as his group of interest.

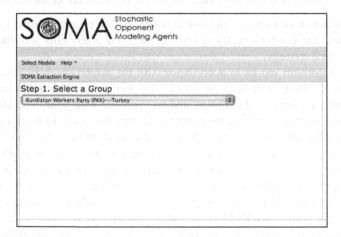

Figure 1: The STOP user has selected the PKK for analysis.

Phase 2. Selecting an Action. Once the user selects a group, a list of actions that this group has engaged in during the period of study (typically the last 25 years or a bit less, depending on how long the group has been in existence) pops up. The user can select one or more actions to study. Figure 2 shows this screen when the user has selected the action "Arms trafficking." Note that there is a number next to each action. This number shows the number of rules we have for the group for this action. For instance, Figure 2 shows that we have 28 rules about "Armed Attacks" for the PKK.

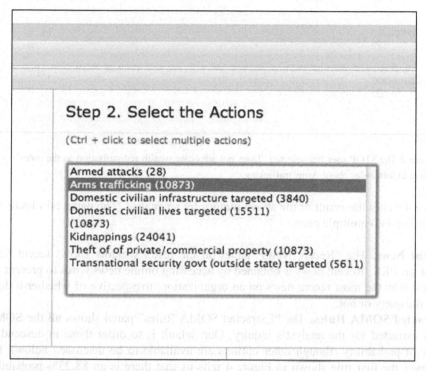

Step 2. Select the Actions

(Ctrl + click to select multiple actions)

Armed attacks (28)
Arms trafficking (10873)
Domestic civilian infrastructure targeted (3840)
Domestic civilian lives targeted (15511)
(10873)
Kidnappings (24041)
Theft of of private/commercial property (10873)
Transnational security govt (outside state) targeted (5611)

Figure 1 The STOP user has selected the action Arms trafficking. There are 10,873 rules about "Arms trafficking" in the system.

Phase 3. Selecting Conditions. Once the user selects an action(s), the system tries to find all rules extracted by SEE about the group dealing with the action(s) selected in Phase 2. Phase 3 allows the user to further focus on those SOMA rules that have certain antecedents. Figure 3 shows what happens when we continue looking at the PKK using the action "Arms trafficking" and we select the condition "Does not advocate wealth redistribution". This selection basically says: Find all rules about the PKK that SEE has derived which deal with the action *arms trafficking* where one of the antecedents of the rule is *does not advocate wealth redistribution*. Once this query has been formulated and submitted, the system computes a web page about the PKK.

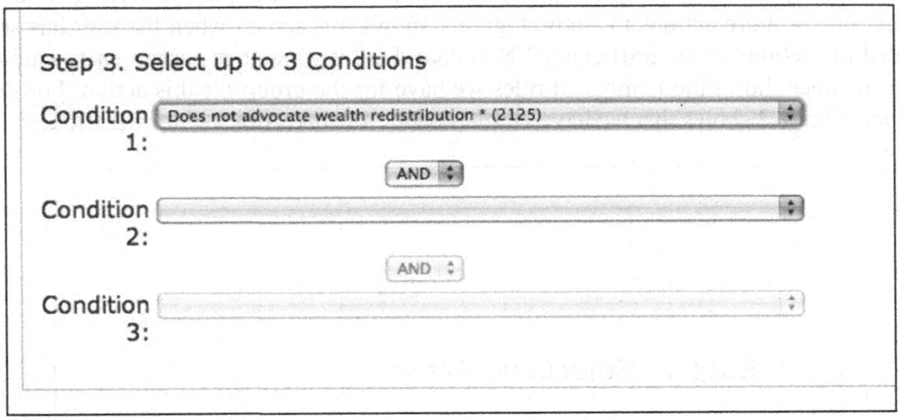

Figure 2 The STOP user has selected "Does not advocate wealth redistribution as the antecedent in their rules about Arms trafficking.

Figure 4 shows the result of the above inquiry by an analyst. The panel on Figure 4 is divided up into multiple parts.

Recent News. The "Recent News" panel on the result window shows recent news about the PKK. Recent news is obtained by accessing online news wires to present the analyst with the most recent news on an organization, irrespective of whether it deals with the query or not.

Extracted SOMA Rules. The "Extracted SOMA Rules" panel shows all the SOMA rules extracted for the analyst's inquiry. Our default is to order these in descending order of probability (though other options are available to be discussed below). For instance, the first rule shown in Figure 4 tells us that there is an 88.23% probability that the PKK engages in the *arms trafficking* action when:

- They do not advocate wealth redistribution and
- They receive foreign state support and
- There is no intra-organizational conflict.

The bottom right of the screen shows that there were a total of 82 SOMA rules we derived for this query and that the first 25 are shown on this screen. The user can browse all these 25 rules on this screen or move to another screen by transitioning to another page (see bottom left of the panel for how to move from one page to another).

Basic Information about the Group. The bottom of the page also shows that STOP can access other sources such as Wikipedia in order to present the user some background information about the group. Sources other than Wikipedia are also planned for access in coming months

Access to other Systems. Other systems such as T-REX [10] can provide valuable information about activities that different groups around the world might be

Figure 3. The results from a query in the STOP system

engaging in. For example, the T-REX system can automatically extract which violent activities groups are involved in. T-REX processes around 150K pages each day from approximately 145 news sources worldwide.

In addition, the CONVEX system [12] under development at the University of Maryland provides an *alternative to SOMA* for forecasting. In a future version of STOP, we anticipate providing access to CONVEX through the SOMA Terror Organization Portal. We also hope to provide access to a wide variety of third party systems as well, but this is not the case at this point.

SANE contains facilities for social networking amongst national security analysts and researchers working on counter-terrorism efforts worldwide. The kind of social networking supported by SANE is shown at the bottom on Figure 4.

- **Mark as Useful.** Analysts can mark certain rules as "Useful". When an analyst marks certain rules as useful, a tuple gets stored in our SOMA Utility Database underlying STOP. This tuple identifies the group, the id of the rule, and the userid of the user who found the rule to be useful. In addition, the analyst may enter a comment about why he or she found the rule useful. This too is stored. Additional metadata about the transaction is also stored.

- **Mark as Not Useful**. Just as in the preceding case, analysts can mark rules as "Not Useful" as well.
- **Show Ratings**. SANE typically shows rules to an analyst in descending order of probability. However the design of **SANE** also allows a "Show Ratings" capability where the analyst can see ratings that other analysts have given to rules. We expect, in the very near future, to allow users to query the SOMA Utility Database so that he can see which rules have been marked as interesting or not interesting by colleagues' whose opinion he respects.

The "Yearly Data" tab towards the bottom left of the screen shows yearly data (i.e. the actual Minorities at Risk Organizational Behavior codes for a given year). The SANE design allows an analyst to pull these up, mark them as useful or not, and also insert comments on the codes. These features, which are in the process of being incorporated into the system, will, upon completion, allow further social networking activities along the lines mentioned above.

4 SOMA Analyst Forecast Engine (SAFE)

The "Forecast" tab towards the bottom left of Figure 4 allows an analyst to invoke the SOMA Analyst Forecast Engine (SAFE). The basic idea in SAFE is to allow an analyst to hypothesize a "state" S of the world and to ask the system what *set* of actions the group G he is studying will engage in.

SAFE uses the technology described in [2,3] in order to achieve this. Based on the state S that the analyst specifies, SAFE focuses on those rules SEE has extracted about group G that are applicable w.r.t. situation S. In other words, SAFE focuses on the set { r | r is a rule SEE has extracted about group G and S satisfies the antecedent of rule r}. This set of rules is called the set of *applicable rules for G w.r.t. state S*, denoted *App(G,S)*.

SAFE then sets up a set *LC(G,S)* of linear constraints associated with G and S as described in [2,3]. Each set of actions that group G can engage in is termed a "world" (or a possible world). The variables in *LC(G,S)* correspond to the (as yet unknown) probabilities of these worlds. By solving one linear program for each world using the algorithms in [2,3], SAFE is able to find the k most probable worlds.

Thus, an analyst can use SAFE in order to hypothesize a state S and to see what the k most probable worlds are. The algorithms underlying SAFE have been fully implemented – at the time of writing of this article, they are being "plugged into" SAFE to support system usage.

We are currently inserting Social Networking features into the SAFE component. When an analyst chooses to do so, he can save a state he has hypothesized, save a prediction he has hypothesized, choose whether he agrees with the forecast or not, and comment on the prediction as well. All these are saved in a *Forecast and Comment* database. Other analysts who are authorized to do so may comment on those same predictions as well. The FAC database will provide a query facility to other users who can see which forecasts when authorized to do so.

5 Conclusions

In this paper, we have described the SOMA Terror Organization Portal (STOP) – a facility that national security analysts can use in order to understand terror threats worldwide. STOP provides a single, Internet or Intranet accessible (password protected) site within which national security analysts can study certain groups. Not only does STOP provide tools they might use, it also provides a valuable social networking capability that allows analysts to often create and expand a network of experts on a given topic. It hardly needs to be said that STOP allows them to leverage this network so that different points of view can be incorporated into their analytic task before a final recommendation is made.

To date, STOP allows analysts the ability to examine approximately 23 terror groups form about ten countries ranging from Morocco all the way to Afghanistan. Users can see rules extracted automatically about these groups (over 14,000 for Hezbollah; over 20,000 for the PKK), browse them, experiment with them, and mark them as useful or not. They can use these markings to build consensus (or at least identify different camps) about a given topic, and explore the pros and cons of alternative views – all without having to move from their desk.

As the powerful real-time capabilities of the T-REX Information Extractor become available to STOP, the need for manually coded information currently used by T-REX will disappear. We expect this transition to start in mid-2008.

Acknowledgements The authors gratefully acknowledge funding support for this work provided by the Air Force Office of Scientific Research through the Laboratory for Computational Cultural Dynamics (LCCD) grants AFOSR grants FA95500610405 and FA95500510298, the Army Research Office under grant DAAD190310202, and the National Science Foundation under grant 0540216. Any opinions, findings or recommendations in this document are those of the authors and do not necessarily reflect the views of sponsors.

References

1. Gerardo I. Simari, Amy Sliva, Dana S. Nau, V. S. Subrahmanian: A stochastic language for modeling opponent agents. Proc. AAMAS 2006, pages 244-246.
2. S. Khuller, V. Martinez, D. Nau, G. Simari, A. Sliva, V.S. Subrahmanian. Action Probabilistic Logic Programs, accepted for publication in *Annals of Mathematics and Artificial Intelligence*, 2007.
3. S. Khuller, V. Martinez, D. Nau, G. Simari, A. Sliva, V.S. Subrahmanian. Finding Most Probable Worlds of Logic Programs, Proc. 2007 Intl. Conf. on Scalable Uncertainty Management, Springer Lecture Notes in Computer Science Vol. 3442, pages 45-59, 2007.
4. A. Sliva, V. Martinez, G. Simari and V.S. Subrahmanian. SOMA Models of the Behaviors of Stakeholders in the Afghan Drug Economy: A Preliminary Report, Proc. 2007 Intl. Conf. on Computational Cultural Dynamics, pages 78-86, AAAI Press.
5. V.S. Subrahmanian. Cultural Modeling in Real-Time, *Science*, Vol. 317, Nr. 5844, pages 1509-1510.
6. A. Mannes, M. Michael, A. Pate, A. Sliva, V.S. Subrahmanian and J. Wilkenfeld. Stochastic Opponent Modeling Agents: A Case Study with Hezbollah, Proc. Intl. Conf. on Social and Behavioral Processing, Sedona, Arizona, April 2008, Springer Verlag Lectures Notes in Computer Science, to appear.
7. V. Asal, C. Johnson, J. Wilkenfeld. "Ethnopolitical Violence and Terrorism in the Middle East" in J.J. Hewitt, J. Wilkenfeld, T.R. Gurr (eds) *Peace and Conflict 2008*. Boulder, CO: Paradigm Publishers. 2007.
8. Minorities at Risk Project. College Park, MD: Center for International Development and Conflict Management. 2005 Retrieved from http://www.cidcm.umd.edu/mar/.
9. V.S. Subrahmanian, M. Albanese, V. Martinez, D. Nau, D. Reforgiato, G. I. Simari, A. Sliva and J. Wilkenfeld. CARA: A Cultural Reasoning Architecture, IEEE Intelligent Systems, Vol. 22, Nr. 2, pages 12-16, March/April 2007.
10. M. Albanese and V.S. Subrahmanian. T-REX: A System for Automated Cultural Information Extraction, Proc. 2007 Intl. Conf. on Computational Cultural Dynamics, pages 2-8, AAAI Press, Aug. 2007.
11. C. Cesarano, B. Dorr, A. Picariello, D. Reforgiato, A. Sagoff and V.S. Subrahmanian. Opinion Analysis in Document Databases, Proc. AAAI Spring Symposium on Computational Approaches to Analyzing Weblogs, Stanford, CA, March 2006.
12. V. Martinez, G. I. Simari, A. Sliva and V.S. Subrahmanian. CONVEX: Context Vectors as a Paradigm for Learning Group Behaviors based on Similarity, in preparation.

The Sociology of Alien Rule

Michael Hechter[†]

[†]michael.hechter@asu.edu, School of Global Studies, Arizona State University, Tempe, AZ

Abstract Discontent with alien rule is often assumed to be pervasive, if not universal, thus accounting for the absence of an international market in governance services. There is no shortage of explanations of the antipathy to alien rule, and a great deal of corroborative evidence. Many believe that people seem to prefer to be badly ruled by their own kind than better ruled by aliens. Yet if this is true, then identity trumps competence in the assessment of rule, implying that we are all liable to suffer from suboptimal governance. In contrast, this paper argues that the evidence for the pervasiveness of antipathy to alien rule is overdrawn. To that end, it distinguishes between two different types of alien rule -- elected and imposed – provides a brief portrait of each, and suggests that when aliens are confronted with incentives to rule fairly and efficiently, they can gain legitimacy even when they have been imposed.

> Actions are held to be good or bad, not on their own merits, but according to who does them, and there is almost no kind of outrage – torture, the use of hostages, forced labour, mass deportations, imprisonment without trial, forgery, assassination, the bombing of civilians – which does not change its moral colour when it is committed by 'our' side.
>
> -- George Orwell

A host of examples is seemingly consistent with Orwell's ironic observation. Among the most important is nationalism, dating at least from the French Revolution, which is a pervasive force in the contemporary world (Hechter 2000). Beyond this, however, antipathy to alien rule is also found at all levels of social organization. At the most intimate level, Cinderella's resentment of her evil stepmother is echoed in empirical research on high levels of the abuse of stepchildren. The employees of firms often rail against the prospect of hostile takeovers, and one of the greatest fears of the members of academic departments is the threat of receivership.

Naturally, to account for such a pervasive phenomenon a host of explanations has been offered. Evolutionary theorists might ascribe the failure of alien rule to the gap in

19

genetic relatedness between aliens and natives, sociologists ascribe homophily (the tendency for like to associate with like) to common values or, alternatively, structural positions, economists highlight the informational liabilities of top-down rule, and historians ascribe the deleterious effects of colonialism to the non-accountability of alien rulers.

If antipathy to alien rule is as universal as these explanations suggest, however, this is troubling because we assume that in modern society people are matched with jobs – particularly highly skilled ones – on the basis of their *competence* rather than their *identity*. Few of us would select a neurosurgeon on the basis of her ethnicity or religion. Moreover, the leaders of professional sports teams, university departments, and capitalist firms often hail from alien cultures, and many of these organizations are frankly multicultural in composition, as well.

Yet if Orwell is correct, then when it comes to *governance*, this modern idea that competence trumps identity doesn't hold. Nowadays native rulers are always preferred to their alien counterparts. In the wake of the massive post- World War II decolonization movements, detailing the many failings of alien rule turned into a booming academic industry. This industry profited from a great deal of empirical support: it was embarrassingly easy to find examples of colonists treating natives, their cultures and lands in the most egregious fashion. At the same time, self-determination – the antithesis of alien rule -- is often credited with the rapid development of the Southeast Asian Tigers and of Ireland, among other former colonies. All in all, alien rule seemed to be a dead letter.

But is it really? After all, some modern instances of alien rule have met with notable success. The American occupation of postwar Japan and the Allied occupation of Germany paved the way for two of the leading states in the world today. The Marshall Plan, which could be viewed as alien rule lite, is commonly regarded as a crowning achievement of modern American foreign policy. By all accounts, some of the United Nations peacekeeping operations have succeeded. And admission to the European Union – a somewhat ineffective ruler housed in Brussels, but an alien one just the same – is eagerly sought by many from Eastern Europe to Turkey.

Evidently, self-determination has not worked its miracles in many other parts of the world. Robert Mugabe may be native to Zimbabwe, but his country is increasingly a shambles. Alexander Lukashenko uses national sovereignty to justify his iron-fisted rule in Belarus. As any schoolchild knows, the list of egregious native rulers in the contemporary world is depressingly long. The *absence* of alien rule has even come under fire: consider the pervasive criticism of the failure of Western (that is, alien) powers to intervene in the Rwandan genocide and Darfur. Indeed, the imposition of alien rule, in the form of neo-trusteeship, has been advocated as a solution to the growing problem of failed states and the threat they pose to international order. Some recent analysts have claimed that by offering access to modern institutions and technology, colonialism often provided the colonies with net benefits. These considerations suggest that any blanket condemnation of alien rule is likely to mislead.

The evidentiary base for a sociology of alien rule lies in history, especially in comparative history. There are large literatures devoted to the history of occupation regimes in particular countries, but very few studies that generalize across these countries' histories. I have looked in vain for an analysis of the Nazi occupation of Europe. There are plenty of studies of the Nazi occupation of France, of the Netherlands, even of the Channel Islands, but I could find hardly any that deal with all of occupied Europe. The reasons are not hard to find. On the one hand, Nazi policy varied not only from country to country but also from month to month and day to day as the course of the war shifted. On the other, the prospects for native resistance to Nazi rule also varied in these ways. In general, resistance was harder to organize in ethnically or religiously divided countries – like Belgium – than in more homogeneous ones -- like France. And the Allied contributions to the resistance, a key resource for insurgents, varied across places and times.

Generalizations about occupation are problematic because each country's experience is likely to be so complex, so contentious and so historically specific that it is difficult to draw any wider lessons from them. To take just one of a multitude of examples, consider the troubled history of Galicia and the Ukraine from 1914 to 1918. These were religiously, ethnically and politically divided territories subject to the rule of at least two imperial powers – Austria-Hungary and Russia -- in the course of a brutal and chaotic world war. The conflicts went far deeper than this, however. Each of the imperial bureaucracies were rent by conflict, as well. We are invariably reminded of the struggles of Powell and Rice against Rumsfeld and Cheney in the first term of the Bush Administration. Similar disputes have been taking place today in the West Wing about a prospective American military intervention in Iran.

The events surrounding alien rule are so complex, so contentious, and so historically specific that it is difficult to draw any wider lessons from them. Despite all of the complexity and contingency of occupation regimes, a sociology of alien rule may indeed be possible. Understanding the conditions for the legitimation of alien rule is the central question of this sociology.

The leading hypothesis, I would contend, is that alien rule can be legitimated if it provides fair and effective governance. I have elaborated on this argument elsewhere (Hechter 2008). Here, I sketch two brief examples of the legitimation of alien rule. The first example comes from 13th century Genoese Republic, in which alien rule was *elected*, or selected, by those who then become subject to it. Like many other Italian city-states (think of Shakespeare's *Romeo and Juliet*, with its portrait of clashes between Montagues and Capulets in an imagined Verona), 13th century Genoa was rent apart by internecine clan conflict. Under an institution known as the *podesteria*, the alien ruler (the *podestà*) was installed by the Genoese to suppress years of spiraling interclan rivalry. The Genoese installed their own alien ruler to forestall a counterpart who died in the first year of his office installed by the Holy Roman Emperor. Because they chose the institution themselves – albeit under an external threat -- they designed it so as to maximize the neutrality and minimize the power of the alien ruler. The *po-*

destàs key task was to avoid interclan conflict and promote social order and economic development in the city state.

The *podestà* was a non-Genoese noble, given governance powers for the period of a year and supported by twenty soldiers, two judges, and servants he brought with him. The *podestà* and his retinue were required to exit the city at the end of his term and not return for several years, and his son could not replace him in office. He was offered very high wages and a bonus if there was no civil war when he left office.

To minimize his incentive to collude with a given clan, he was selected by a council whose members were chosen on a geographic basis (to prevent its control by any one clan). He – and all relatives to the third degree – was prohibited from socializing with the Genoese, buying property, getting married, or managing any commercial transactions for himself or others. Until permanent housing was built, he had to move his residence around the city – to prevent him from associating for too long with the members of any particular clan.

Moreover, the system provided an administrator controlling Genoa's finances, which limited the clans' ability to expropriate income as a means of increasing their military power. We can infer that the design was successful for the institution lasted for 150 years. Under these highly particular conditions elected alien rule can rather easily become legitimate.

The second example of alien rule comes from 19th and 20th century China, in which rule was *imposed* by forces external to the society. The Chinese Maritime Customs Service (CMCS) was an international, predominantly British-staffed bureaucracy under the control of successive Chinese central governments from its founding in 1854 until 1950. Like the Genoese, the Qing Empire also found itself threatened by foreign powers. For centuries, the center had maintained order by rotating its agents among the various provincial units. In the early part of the 19th century, the western powers eagerly sought export markets for their manufactured goods and aggressively pursued free trade with China. Since the Chinese economy was largely self-sufficient, there had traditionally been little domestic interest in foreign trade. All that changed, however, with the British-instigated importation of opium -- an import that created its own demand. The vast amounts of money lost to the opium trade weakened the Qing regime and helped to fund the Taiping rebels. Just as in Genoa, social order was breaking down in Kwangtung, alarming both the imperialist powers and the Qing government.

Following China's military defeat in the Opium Wars the Qing were induced to accept a series of treaties that eroded aspects of Chinese sovereignty at the point of a gun. The CMCS was a by-product of these treaties, providing alien control over the maritime customs service, which was responsible for collecting revenue derived from overseas trade. It soon came to provide a panoply of public goods, including domestic customs administration, postal administration, harbor and waterway management, weather reporting, and anti-smuggling operations. It mapped, lit, and policed the China coast and the Yangzi. It was involved in loan negotiations, currency reform, and financial and economic management. It even developed its own military forces. In time, the CMCS evolved into one of the most effective parts of the Qing government, thriving even after the nationalist revolution of 1912. Both of these examples of alien rule survived long after the initial conditions for their emergence. This suggests that each attained at least some measure of legitimacy.

But the willingness of alien rulers to govern well hinges on their incentives to do so. Given the appropriate incentives, alien rulers can be motivated to provide fair and effective governance – and, hence, to earn some measure of legitimacy. Alien rulers can be motivated to provide fair and effective governance when aliens impose rule on native territory in order to augment their own security (for which they are prepared to incur some cost), or share in the profits of increased trade and commercial activity. The Genoese were able to design an institution that tied the *podestàs* compensation to his ability to secure social order in the Republic. Likewise, it was in the vital interest of the CMCS to prop up a shaky Qing regime so as to profit from favorable foreign trade policies that the Qing had signed with the Western powers. This motivated the British Inspector Generals of the CMCS to exercise fair and effective rule.

Fair and effective rule, in turn, rest in good part on the treatment of native intermediaries who serve as brokers between rulers and ruled.. Consider the Nazi occupation of Western Europe, one of the most brutal examples of military occupation in modern history. As Robert Paxton (2001: xiii) reminds us, when Hitler conquered France, he wished

to spend as little as possible on the occupation of France. He did not station large military and police forces there after the combat units prepared for Operation Sea-lion against England were withdrawn in December 1940. Many of the remaining able-bodied men were drained away to the Russian front in the summer of 1941, leaving only sixty Home Guard battalions (30,000-40,000 overage men, mostly unfit for combat).

To the degree that alien rulers are resented by native populations, their costs of control must be correspondingly higher than those of native rulers. How can these surplus costs ever be borne? The answer lies in the use of native intermediaries who are induced to collaborate with the alien power to govern the native population. Collaborators are essential because occupation regimes, like their colonial counterparts, always aim to rule on the cheap. Since they depend on collaborators to provide social order, which is the prerequisite for any of their aims in occupying the country in the first place, it is not in the interest of alien rulers to undercut their authority.

Does the antipathy to alien rule mean that we are condemned to worse rule than might be obtained if there were an international market in governance services? Not necessarily. An international market in governance services would offer alien rulers ample incentives to provide fair and effective governance. The role of competitive markets in disciplining the producers of goods and services is powerful. Moreover, the effects of market discipline are culturally universal. Contemporary exemplars of successful alien rule are likely to hasten the advent of such an international market. This is good news for those seeking to improve the performance of governments in an era in which cultural politics has attained exceptional salience.

References

1. Hechter, Michael. 2000. Containing Nationalism. Oxford; New York: Oxford University Press.
2. Hechter, Michael. 2008. Alien Rule and Its Discontents. American Behavioral Scientist Forthcoming.
3. Paxton, Robert. 2001. Vichy France: old guard and new order 1940-1944. New York: Columbia University Press.

The DoD Encounters the Blogosphere

Rebecca Goolsby

rebecca.goolsby@navy.mil, Office of Naval Research, Arlington, VA

Abstract Social computing is exploding and the imagination of the Department of Defense is overflowing with ways to exploit this brave new world. Do jihadists have Facebooks? Can "we" (the good guys) use our technological genius to discover and surveil "them" (the bad guys) as they express themselves and seek to find an audience in blogs, forums, and other social media? Can "we" find the bad guys? This electronic spygame excites some, while other reject it as unworkable, undoable, and maybe not even desirable. Certainly the question of who "we" are is problematic.

Blogs and related social media provide windows into culture, insights into social processes, opinion and political activity, that much is defensible. In places like Iran, where Westerners have few direct contacts or opportunities to interact with the local people, blogs provide critical understandings of how thousands of people construe the world. The blogs themselves are implicitly biased samples, but if one can account for those biases, one can glean an improved awareness. Conceivably, "adversarial" blogs can enrich our understanding of goals, objectives, and reasoning of those who might support anti-Western violence either implicitly or actively.

The Department of Defense is very aware of blogs. They are consummate bloggers, with hundreds of blogs on the two highly classified networks that serve this community (JWICS and SIPRNet). Like other Westerners, they frequently read commercial and media blogs, such as Slashdot, Slate, Wired, and even Wonkette. They imagine the "adversary" has a mirror image "cyber-underworld," where there are equivalent facilities. Is there an easy way to find cyber-underworld that is serving as a mass recruitment source? What does this cyber underworld really look like? What do we know about it and what do we need to know to begin to put together the right questions? Is "targeteering" a feasible goal? What sort of research into blogging and social media should the DoD support in order to diminish violent behavior? In order to better understand the social worlds in which the DoD must accomplish their missions? What kinds of boundaries and safeguards should be put around this type of research in order to ensure ethical research behavior? Who should be involved in developing a research agenda for social computing, national security and defense?

Integrating Multi-Agent Technology with Cognitive Modeling to Develop an Insurgency Information Framework (IIF)

LeeRoy Bronner[†] and Akeila Richards[*]

[†] jlbronner1@verizon.net, Morgan State University, Baltimore, MD
[*] arichards@hotmail.com, Morgan State University, Baltimore, MD

Abstract This research focuses on the application of multi-agent technology and cognitive modeling to develop a decision model for use in analyzing and evaluating behavior strategies of insurgents in Iraq. Insurgency is a complex behavioral process that has many facets. It is a social as well as military problem and a major cause of injury and death for the citizens and military personnel in Iraq. Social computing is concerned with the study of social behavior abstracted by computational models. Computational modeling provides for social system analysis and evaluation. Social computing can be represented as 3-D simulations of real-world events to train, educate or aid in decision making. Users can maneuver through the life-like 3-D Virtual World to observe, record, and measure various behaviors of agents in relation to their environment. The methodology used in modeling these social phenomena is Agent-Oriented Programming (AOP) which is characterized by the use of real-world objects for design and an agent oriented language for implementation. This research integrates the development of multi-agent societies, Electronic Institutions, and Virtual World technology to conduct social computational modeling.

1 Introduction

1.1 Terms

Insurgency: is an armed rebellion by any irregular armed force that rises up against an established authority, government, or administration.
Terrorism: is the systematic use of terror, manifesting itself in violence and intimidation for the purpose of creating fear in order to achieve a goal.

<u>Framework:</u> a logical structure for classifying and organizing information.

<u>Model:</u> a general framework for the representation of a system that allows for reasoning about and investigation of the properties of the system.

<u>Architecture:</u> is the specification of the parts and relationships among these parts of a system and the rules for the interactions within these relationships.

<u>Platform:</u> a set of subsystems and technologies that provide a coherent set of functionality through interfaces and usage patterns. Examples are operating systems, programming languages, databases, and middleware [1].

<u>Model Driven Architecture (MDA):</u> is a conceptual framework and a set of standards to express models, model relationships, model-to-model transformations supported by automated tools and services [1].

<u>Software Agent:</u> a piece of software which performs a given task using information gleaned from its environment. The software should be able to adapt itself based on changes occurring in the environment. Also, an agent's state is viewed as consisting of mental entities (e.g., beliefs, capabilities, choices, commitments).

<u>Agent Oriented Programming (AOP):</u> is a specialization of the Object-Oriented Programming (OOP) paradigm. However, AOP supports a societal view of computation.

<u>Agent Middleware:</u> is the software that provides the framework for enabling easier and more effective agent development by providing general services through high level libraries.

<u>Electronic Institution (EI):</u> an infrastructure that establishes a set of norms or rules on the behavior of participants that can be humans or autonomous agents.

<u>Unified Modeling Language (UML):</u> is a graphical representation language of symbols that allows analysts to develop several different types of visual diagrams that abstracts various aspects of a system.

<u>Enterprise Architect (EA):</u> is a comprehensive UML analysis and design tool, covering system development from requirements gathering, through to the analysis stages, design , testing and maintenance.

<u>Executable Unified Modeling Language (xUML):</u> software that allows a developer to define the behavior of an object-oriented model in sufficient detail that it can be executed. The executable model describes the data and behavior in classes depicting the application [2].

1.2 Purpose of Research

Advanced software engineering technology has been used in past decades to solve a host of global issues from traveling to the moon to exploring under seas. There lacks, however, similar sophisticated means to solve crucial problems concerning sociological systems. For example, the US Army is asking more questions concerning social phenomena as a result of the escalated death tolls in the Iraq War. When the US Army is deployed in foreign territories, determining the psychosocial milieu can be as important as determining that equipment is available and troops are combat ready. In an environment of information overload about Iraq and the insurgency problem in particular, it is difficult to extract significant information when it is not specifically tailored for military use. Therefore, how can insurgency information best be researched, defined, modeled and presented for more informed decision making?

The purpose of this research is to provide the military with a framework and tools that produce information with which to better understand insurgency and behavior of the insurgent. This research addresses insurgency as a social computing problem, which will facilitate behavioral modeling. Social computing is concerned with the study of social behavior and social context based on computational systems. Behavioral modeling reproduces the social behavior, and allows for experimenting, scenario planning, and understanding of behavior, patterns and potential outcomes. Research of this type must be done in collaboration with researchers from multiple disciplines. Information and Civil Engineering along with the Psychology department at Morgan State University (MSU) have undertaken this project as a collective effort.

Social computing has been a way to structure formal modeling for years; however, the complexity of modeling tools has posed difficulty in developing models. A social computing model has been researched across the past two years to address insurgency in Iraq where the most difficult part of this effort is the computational abstraction of the insurgent's behavior. A number of cognitive modeling tools (e.g., Atomic Components of Though- Rational (ACT-R) [3], Cougaar [4]) have been used in the past. However, the process has been slow and difficult. Currently a third modeling process and methodology is being researched. The tool is Jadex (Java Agent Development Framework Extension) [5] and the behavioral model being developed is based on the Belief-Desire-Intention (BDI) [6] Paradigm. BDI is a well-known model for representing mental concepts. The computational abstractions developed in this research are being modeled based upon the Model Driven Architecture (MDA) [1] paradigm, Multi-Agent Systems (MAS) [7], Electronic Institutions (EI) [8,13], the BDI Paradigm and visualization of solutions through Virtual World technology. . The product of this research effort will be an Insurgency Information Framework (IIF) model which can be used to characterize insurgency in Iraq.

2 Research System

This research addresses human behavior, complex system modeling and graphical information systems. It was necessary to develop a research team (figure 1) with researchers from Psychology, Information Systems, Systems and Civil engineering disciplines. Through this collaboration the Insurgency Information Framework

Figure 1: Insurgency Research System

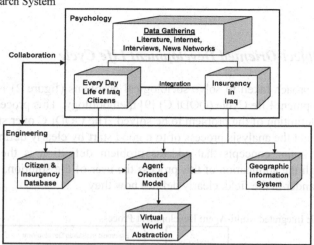

model is being developed. The Psychology department has the responsibility for collecting data and developing behavioral concepts to drive the IIF. Much of the data is being collected from the internet through articles, newspapers, journalist sites, and blogs. Also, interviews have been scheduled with other knowledgeable people. It has been quite difficult to acquire the kind of data the Army is interested in due to security classifications. However, at a workshop in which the Army played a prominent role, it was brought out that gang activity in our urban cities can in some ways approximate insurgency activity in foreign regions. Where breaking the law (e.g., breaking and entering, assault, etc.) in certain locations in the city can be associated with armed military activity such as setting and detonating an Improvised Explosive Device (IED).

As shown in figure 1, the Psychology researchers will prepare and submit their data to engineering for database development and the development of behavioral models. Also, data will be used to develop Geographic Information System (GIS) maps. These GIS maps will be integrated into the overall modeling effort to be invoked where necessary to aid commanders in their decision making process. The agent oriented model developed from this data will be used to drive the Virtual World abstraction of the model solution. Again, the goal of this research is to bring our solutions to life by making clear visual representations of the behavior that is being abstracted via avatars in a 3D Virtual World setting.

3 Methodology

3.1 Object-Oriented Development Life Cycle

The approach taken to solve sociological problems (figure 2) is the Object-Oriented Development Life Cycle (OODLC) [9] methodology. This process begins with a thorough definition of the problem to be solved. The OODLC over stresses how important it is to get the analysis process of to a good start by clearly defining the problem. One of the major concepts that address problem definition is the Use Case Analysis. Through the application of this process the user of the solution, in this case the Army commander in the field, clearly defines how they

Figure 2: Integrated Multi-Agent Development Process

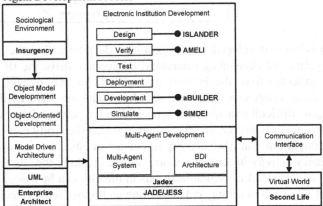

plan to use the system. At problem definition time an object-oriented model is developed. The real beauty of the OODLC process is that the same model developed during problem definition is carried throughout all phases of the process. After problem definition comes the analysis, design, implementation and evaluation phases.

The purpose of the OODLC in the overall methodology is to develop the initial model to be used with the Agent-Oriented development process. In this methodology, the concepts of the Model Driven Architect (MDA) approach to analysis are superimposed on the OODLC methodology to enhance the model development process. The MDA approach applies more standards to the modeling process and provides specific tools to automate the process. Greater standardization makes it possible to design automating tools for the development process. There are three steps to the MDA modeling process. These are the development of the Computational Independent Model

(CIM), development of the Platform Independent Model (PIM), and the development of the Platform Specific Model (PSM). The CIM focuses on the context and requirements of the system without the consideration for its structure or processing. It defines what the system should do not how it should do it. The PIM describes the systems in terms of certain analysis artifacts (e.g., classes, objects, attributes, etc.) but does not describe any platform-specific choices. The PSM is the PIM where attention is paid to the specific details relating to the use of a particular platform. In this methodology, the Enterprise Architect (EA) [1,10] tool is being used. EA is a comprehensive UML analysis, design and maintenance tool. With the EA it is possible to develop an artifact database which can be indexed, accessed and managed.

3.2 Multi-Agent Development

An agent is a computational piece of software developed using a language such as Java supported by agent middleware such as (Java Agent Development Framework (JADE) [15,17]. JADE is an enabling piece of software designed for the development and run-time execution of applications based on the agent paradigm. The agent paradigm abstracts a model that is autonomous, proactive and social. Multiple agents can be abstracted on a JADE platform. JADE makes it possible for agent interaction through discovery and communication (i.e., communication concepts are derived from speech act theory [11]) by exchanging asynchronous messages.

The Java Expert System Shell (JESS) [16,17] is a rule-based system that has the capacity to reason using knowledge in the form of declarative rules. Rules are a lot like "if-then" statements of traditional programming languages. A rule-based system is a system that uses rules to derive conclusion. JESS is being used as a part of this methodology to provide rule-based reasoning to support the rules and policies implemented within the Electronic Institution Paradigm. The JESS engine will be integrated with the JADE agent to provide the capability for specification of institutional rules.

Jadex is an add-on to the JADE agent platform. The add-on follows the Belief-Desire-Intentions (BDI) paradigm, a well known model for representing mental concepts in the system design and implementation of agent systems. The objective of this add-on is to facilitate the utilization of mental concepts in the implementation of agent environments.

3.3 Electronic Institutions

Electronic Institutions are the models that define the rules for agent societies by fixing what agents are permitted and forbidden to do and under what circumstances. The basis for the development of Electronic Institutions in this methodology is the 3D Electronic Institutions [8,12] paradigm. The 3D Electronic Institution model proposes viewing Virtual Worlds as open Multi-Agent Systems. 3D Electronic Institutions are Virtual Worlds with rules and policies that regulate agent interactions. The methodology proposed by 3D Electronic Institutions separates the development of Virtual Worlds into to two independent phases: specification of institutional rules and design of the 3D Virtual World agent interaction.

There are four tools (figure 2) that are utilized to support this methodology:

1. ISLANDER: a graphical tool that supports the specification and static verification of institutional rules.
2. AMELI: a software platform to run Electronic Institutions for verification.
3. aBUILDER: an agent development tool that generates agent skeletons. The generated skeletons can be used either for EI simulations or for actual executions.
4. SIMDEI: a simulation tool to animate and analyze ISLANDER specifications.

3.4 Communication Interface

The purpose of the Communication Interface is to causally connect the institutional infrastructure with the visualization system and transform the actions of the visualization system into the messages, understandable by the institutional infrastructure. Messages created by this interface assist in maintaining synchronization and consistent relationships between the Virtual World and the EI [13].The causal connection is made by a server termed the Causal Connection Server. This casual connection uses an Action-Message table (Table I) [12]. The casual connection happens in the following way: an action executed in the Virtual World (that requires EI verification) results in a change of the institutional state as well as every change of the institutional state is reflected onto the Virtual World and changes its state.

3.5 Virtual Worlds [14] – Second Life

A Virtual World is an interactive simulated environment accessed by multiple users through an online interface. This is the world in which avatars will interact and accomplish the goals of their users. Virtual Worlds are also called "digital worlds," or "simulated worlds." There are commercial Virtual Worlds available today. One of the most well known is Second Life [15]. This research is attempting to make use of Second Life as an example of the integration of the 3D Metaphor with a visualization environment. Virtual Worlds have been created for many different purposes. The following summarizes a few ways in which they are currently used: simulation, education and training, decision making, and socializing.

4 The IIF Model

The Insurgency Information Framework (IIF) model (figure 3) is developed from three sub-models. The Geographic Information System (GIS) model at base level-1 is used to present and analyze geographic information. Using this GIS Model, it is possible to create interactive maps where the user and other sub-models can select hot spots and view insurgency information. The Application Context Model (ACM) abstracts the environment where the application activity will take place. Also, the ACM will implement the Electronic Institution. For example, the ACM can depict the urban environments (e.g., buildings, streets, stores, mosques) where insurgency can take place. The Application Model (AM) and the ACM are implemented as Multi-agent Systems. This allows buildings, stores, and streets to be just as versatile as the avatars.

Table 1. Action-Message [12].

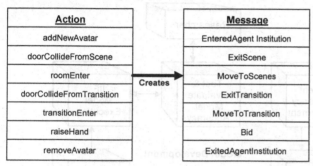

Action		Message
addNewAvatar		EnteredAgent Institution
doorCollideFromScene		ExitScene
roomEnter	Creates	MoveToScenes
doorCollideFromTransition		ExitTransition
transitionEnter		MoveToTransition
raiseHand		Bid
removeAvatar		ExitedAgentInstitution

Figure 3: The IIF Model

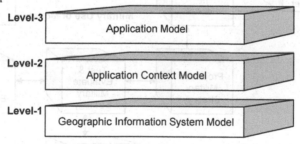

Level-3 Application Model

Level-2 Application Context Model

Level-1 Geographic Information System Model

5 Use of the IIF Model

The Insurgency Information Framework (IIF) Model is developed in two phases. First, the object-oriented model is developed (figure 2). The object-oriented model is input to the Model Driven Architecture (MDA) process. MDA is well suited to be implemented via automated tools. Also, MDA provides the input for the AOP step. As shown in figure 2, JADE and Jadex will be used to develop the agent behavioral models. These agent models will be used to control the avatars in the Virtual World. The initial model will reflect the current insurgency environment (figure 4). The current IIF Model will be tested for quality. Executable Unified Modeling Language (xUML) is a tool used in testing the quality of object-oriented models. xUML fully supports the MDA paradigm. In this methodology, xUML

Figure 4: Use of the IIF Model

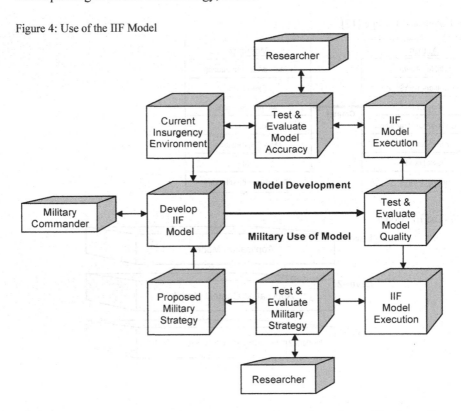

will be used to verify that the model meets object-oriented specifications. For example, does the relationship between classes meet UML specifications, such as class inheritance, association rules, etc.? Accuracy as opposed to quality is addressed by testing how well the IIF model depicts the current insurgency environment.

In the second phase of the model development process, a military strategy to address the insurgency problem is modeled. This strategy is integrated with the model developed in the first phase. As in the initial model development process, the model is evaluated for quality and accuracy. The avatars, as a part of the Application Model, will be integrated with the insurgency ACM to determine how the avatar insurgents will react to the environment created by the military strategy. This will allow the strategy to be evaluated and interpreted before it is used by the military in the field.

6 Summary

This paper addresses modeling and analysis of social systems using a multiple discipline approach. In the introductory section 1, the purpose of this research is outlined. The purpose is to develop an Insurgency Information Framework (IIF) model to aid the military in understanding insurgency and the behavior of insurgents. In section 2, the system used to perform this research is defined. Section 3 presents an approach for developing a methodology to be used for analyzing social systems. This is a complete methodology using object and agent oriented technology, Electronic Institutions, and Virtual World technology. The final sections of this paper defines the IIF Model and how it can be used by the military in developing strategies to address insurgency in Iraq as well as to improve their decision making process. Databases are needed to track entities such as agents, geography, activities, libraries of geometric data, etc. Also, object-oriented based tools are required to manage the large number of artifacts of analysis (e.g., use cases, classes, objects, etc.). This research is in progress and is a continuing effort.

References

1. Truyen Frank (2005) Implementing Model Driven Architecture using Enterprise Architecture. Sparx Systems, Victoria, Australia
   http://www.sparxsystems.com.au/downloads/whitepapers/EA4MDA_White_Paper_Features.pdf
2. Mellor, S J, Balcer, M J (2002) Executable UML: A Foundation for Model-Driven Architecture. Addison-Wesley, New York.
3. Taatgen N et al Modeling Paradigm in ACT-R.
   http://www.ai.rug.nl/~niels/publications/taatgenLebiereAnderson.pdf

4. Belagoda T et al (2005) A Critical Study of Cougaar Agent Architecture.
   http://www.utdallas.edu/~kandula /docs/COUGAAR_final.pdf
5. Braubach L (2003) Jadex: Implementing a BDI-Infrastructure for JADE Agents
   http://vsis-www.informatik.uni-hamburg.de/getDoc.php/publications/124
   /pokahrbraubach2003jadex-xp.pdf
6. Brazier F et al (1996) Beliefs, Intentions and Desires.
   http:// ksi.cpsc.ucalgary.ca/KAW/KAW96/ brazier/ default.html
7. Wooldridge Michael (2002) An Introduction to Multi-Agent Systems. Wiley, New York.
8. Arcos Josep et al (2005) An Integrated Development Environment for Electronic Institutions.
   http://www.iiia.csic.es/~jar/papers/2005/agent-book-camera.pdf
9. Brown D (1997) An Introduction to Object-Oriented Analysis. Wiley, New York.
10. SPARX Systems, Enterprise Architect, Version 7.0,
    http://www.sparxsystems.com.au/products/ea.html
11. Grice P (1989) Studies of the Ways of Words. Harvard University Press, Massachusetts.
12. Bodganovych A et al (2007) A Methodology for Developing MAS as 3D Electronic Institutions.
    http://users.ecs.soton.ac.uk/mml/aose2007/accepted/assets/9-bagdanovych-fin.pdf
13. Bogdanovych Anton et al (2005) Narrowing the Gap between Humans and Agents in e-Commerce: 3D Electronic Institutions.
    http://ispaces.ec3.at/papers/BOGDANOVYCH.EA_2005_narrowing-the-gap.pdf
14. Book Betsy (2004) Moving Beyond the Game: Social Virtual Worlds.
    http://www.virtualworlds review.com/papers/BBook_SoP2.pdf
15. Kapor M (2006) The Power of Second Life.
    http://www.3pointd.com/20060820/mitch- kapor-on-the-power-of-second-life/
15. Bellifemine F et al   (2003) JADE: A White Paper.
    http://jade.tilab.com/papers/2003/WhitePaperJADEEXP.pdf
16. Friedman-Hill E (2003) JESS in Action: Rule-Based Systems in Java. Manning, Connecticut.
17. Cardoso H L (2007) Integrating JADE and JESS.
    http://jade.tilab.com/doc/tutorials/jade-jess/jade_jess.html

Stochastic Opponent Modeling Agents: A Case Study with Hezbollah

Aaron Mannes[1], Mary Michael[2], Amy Pate[3], Amy Sliva[4], V.S. Subrahmanian[5], Jonathan Wilkenfeld[6]

[1]amannes@umd.edu, [2]marym@umd.edu, [3]apate@start.umd.edu, [4]asliva@cs.umd.edu, [5]vs@cs.umd.edu, [6]jwilkenf@gvpt.umd.edu, University of Maryland, College Park, MD 20742, USA.

Abstract Stochastic Opponent Modeling Agents (SOMA) have been proposed as a paradigm for reasoning about cultural groups, terror groups, and other socio-economic-political-military organizations worldwide. In this paper, we describe a case study that shows how SOMA was used to model the behavior of the terrorist organization, Hezbollah. Our team, consisting of a mix of computer scientists, policy experts, and political scientists, were able to understand new facts about Hezbollah of which even seasoned Hezbollah experts may not have been aware. This paper briefly overviews SOMA rules, explains how more than 14,000 SOMA rules for Hezbollah were automatically derived, and then describes a few key findings about Hezbollah, enabled by this framework.

1 Introduction

Stochastic Opponent Modeling Agents introduced in [1,2,3] were introduced as a paradigm for reasoning about any group G in the world, irrespective of whether the group is a terror group, a social organization, a political party, a religious group, a militia, or an economic organization. SOMA-rules have been used to encode the behavior of players in the Afghan drug economy [4] as well as various tribes along the Pakistan-Afghanistan border [5].

In contrast with the above groups, Hezbollah is a well-known terrorist organization based in Lebanon. In September 2002, testifying to Congress, Deputy Secretary of State Richard Armitage stated, "Hezbollah may be the A team of terrorists, and maybe al-Qaeda is actually the B team." [6] Prior to 9/11 Hezbollah was the terrorist organization that had killed the most Americans, carrying out the massive suicide bombings of the U.S. Marine Barracks and U.S. Embassy in Beirut in the early 1980s. Hezbollah also orchestrated a campaign kidnapping Westerners in Lebanon that triggered political crises in the United States and France. Hezbollah terror attacks have, however, extended far beyond the Middle East, to Europe and Latin America.

Ideologically, Hezbollah expounds a radical version of Shia Islam that seeks violent confrontation with enemies of Islam such as the United States and Israel. Rooted in Lebanon's long oppressed Shia community and closely allied with Iran and Syria, Hezbollah views itself as the spearhead in Islam's struggle with the West. At the same time, Hezbollah relies on the support of Lebanon's Shia community and its Iranian and Syrian sponsors and consequently its actions are shaped by these factors.

Hezbollah is also a multi-faceted organization that engages in a range of activities to further its cause. It participates in Lebanese elections and runs businesses and social services. It maintains a guerilla force that fought a multi-year insurgency against Israel in South Lebanon and conducted platoon and company level operations against Israel during the summer of 2006. Internationally, it provides training to Islamist terrorists including al-Qaeda and Palestinian terrorist groups. Because of its perceived success against Israel it has been lionized throughout the Arab world. To burnish its image it runs a satellite television station, radio stations, and even produces a video game. Hezbollah also has links to the Lebanese Shia diaspora, which has a presence on every continent. [7]

Hezbollah's combination of formidable capabilities, radical ideology, and international reach makes developing systems to better understand Hezbollah's operations and, if possible, predict them an important priority.

In this paper, we present an overview of a few of the more than 14,000 rules about Hezbollah's behavior that our SOMA system has extracted *automatically*. Of course, presenting all these rules is impossible in the context of a short paper – hence, we briefly describe our rule derivation methodology and then describe core SOMA results about Hezbollah.

2 SOMA Rule Derivation Methodology

We derived SOMA rules from the *Minorities at Risk Organizational Behavior (MAROB)* dataset [8], which is an extension of the Minorities at Risk (MAR) dataset [9]. MAR tracks the repression, discrimination and political behaviors, such as rebellion and protest, for 284 ethnic groups worldwide. In an effort to better understand the nature of political violence, MAROB was created at the University of Maryland in 2005 to track behaviors and characteristics of ethnopolitical organizations, those claiming to represent MAR ethnic groups. As nine of the 14 most deadly terrorist organizations from 1998 to 2005 were ethnonationalist, MAROB reflects the importance of studying ethnopolitical organizations.

From a computational point of view, MAROB associates a relational database table with each group. The rows of the table reflect different years. The columns of the table denote different properties about the behavior of that group or about the environment within which the group functioned. For instance, a column such as KIDNAP specifies if the group used kidnapping as a strategy in a given year. Likewise, a column named FORSTFINSUP specifies if the organization got financial support from a foreign state during a given year. The columns of any relational database table associated with a MAROB group fall into three categories: columns about *actions* that the group took (such as KIDNAP above), columns about the *environment* in which the group functioned (such as FORSTFINSUP above), and other *administrative* columns. Note that the *environment* can include information about actions that *other* groups took that contributed to the climate in which the group being modeled exists.

Our SOMA rule extraction method used MAROB data from 1982 to 2004 in order to extract rules about Hezbollah. A SOMA rule about a group G has the form

$$\text{<Action>}:[L,U] \; \underline{if} \; \text{<Env-Condition>}$$

Where:

- <Action> is an action that the group took (such as KIDNAP)
- <Env-Condition> is a logical conjunction of elementary conditions on the environmental attributes. An elementary condition associated with the environmental attribute A is an expression of the form $A \; \underline{op} \; value$ where $\underline{op}$ is in the set $\{ =, <=, >= \}$.
- [L,U] is a closed sub-interval of the [0,1] interval.

The above rule says that in any year when the <Env-Condition> is true, there is a probability between L and U that the group took the action stated in the rule. The rule below is an example of a rule that we extracted about Hezbollah.

KIDNAP: [0.51,0.55] $\underline{if}$ solicits-external-support & does not advocate democracy.

This rule says that in years when Hezbollah both solicited external support and did not promote democratic institutions, there was a 51 to 55% probability that they engaged in kidnapping as a strategy.

The SOMA rule extraction method consists of three steps:
1. Select a value for <Action>,
2. Fix one environmental attributes as part of <Env-Condition>,
3. Add varying combinations of up to three of the remaining environmental attributes to <Env-Condition> to determine if significant correlations exist between <Env-Condition> and <Action>.

Using the standard definition of confidence from the literature, the rule extraction method calculates the difference between the confidence value produced by <Env-Condition> and its negation. If this difference exceeds a given threshold, then a SOMA rule is extracted. To obtain the probability range for the extracted rule, we use the confidence value plus/minus ε. This process is repeated for all combinations of environmental attributes and actions.

By analyzing the MAROB data for a period of 23 years, we identified more than 14,000 rules for Hezbollah's behaviors.

3 Some Results about Hezbollah's Behavior

SOMA provided probabilities for four different Hezbollah actions: armed attacks, targeting domestic security forces, kidnappings, and transnational attacks. Due to space constraints, we focus on the rules on kidnappings and transnational attacks. In general, the rules are in accord with understood patterns of Hezbollah activities – while revealing some new insights about the triggers for these activities.

The central condition for the probabilities of kidnapping and for committing transnational attacks (including Katyusha rocket strikes against Israel and external terror attacks such as the 1994 bombing of the Jewish community center in Buenos Aires) is Hezbollah's relationship to Lebanese politics. From 1974 until 1992 Lebanon did not hold elections because of an on-going civil war. Prior to 1992 Hezbollah could not participate in Lebanese elections and did not attempt to represent its interests to Lebanese officials. In 1992 Hezbollah had a strategic shift in its relationship with the traditional Lebanese power structures and began to represent its interests to Lebanese officials by participating in elections. Hezbollah's leadership shifted from seeking to transform Lebanon into an Islamic state to working within the system to pursue its goals. Hezbollah's goals, particularly regarding confronting Israel, and its willingness to turn to violence against enemies both within Lebanon and without did not change. But the tactics did. Prior to Hezbollah's strategic shift, kidnapping was a primary tactic used by Hezbollah to gain stature. With the end of the Lebanese Civil War and Hezbollah's entry into Lebanese politics, the likelihood of kidnapping dropped substantially and the likelihood of committing trans-national attacks increased dramatically.

3.1 Hezbollah's Kidnapping Campaign

Table 1. Conditions and Probabilities for Kidnapping.

Conditions	Probability
Does not advocate democracy & solicits external support	.53
No foreign state political support & major inter-organizational conflict	.53
Solicits external support & does not advocate democracy & no foreign state political support	.66
Major inter-organizational conflict & no foreign political support & (foreign state provided non-violent military support OR standing military wing)	.66
Soliciting external support is a minor strategy & (electoral politics is not a strategy or does not advocate democracy)	.83

The conditions relating to increased probabilities of kidnapping reflect Hezbollah's capabilities and opportunities. Receiving military support or possessing a standing military wing would increase capabilities. Inter-organizational conflict represents an opportunity. Hezbollah and its rival Amal both conducted kidnappings as part of their struggle for primacy among the Lebanese Shia community.

The strongest condition linked to a Hezbollah kidnapping campaign is soliciting external support. When soliciting external support, Hezbollah leaders are meeting with other leaders and the organization is opening offices in other countries. In the Middle East kidnapping campaigns against the West and Israel are useful for raising an organization's profile – thereby making it a more attractive candidate for support. Kidnapping creates bargaining chips. When holding hostages, Hezbollah can either attempt to extract support from the hostages' nation of origin or give potential supporters the opportunity to act as an interlocutor between Hezbollah and the hostages' nation of origin. During the Lebanese Civil War, when Hezbollah efforts to obtain external support were greater, it appears that they were more likely to curtail their kidnapping activity – possibly in response to pressures from potential supporters.

3.2 Hezbollah's Transnational Attacks

Once in Lebanese politics, transnational attacks became the more attractive strategy. Terrorist attacks outside Lebanon could be denied and did not have a substantial impact on Lebanon itself. Since Lebanon does not have relations with Israel and many Lebanese resented Israel's long-standing security zone in the south of Lebanon, Hezbollah's rocket attacks against Israel did not detract from Hezbollah's domestic political standing, and in some cases may have increased it.

Table 2. Conditions and Probabilities for Transnational Attacks

Conditions	Probability
Pro-democracy ideology	.52
Electoral politics is a minor strategy & no foreign political support	.55
Medium inter-organizational conflict	.58
Electoral politics is a minor strategy & no non-military support from the diaspora	.6
Electoral politics is a minor strategy & (medium rioting OR no foreign state political support)	.6
Electoral politics is a minor strategy	.635
Electoral politics is a minor strategy & medium inter-organizational conflict & no foreign state political support	.67
Electoral politics is a minor strategy & medium inter-organizational rioting	.67
Electoral politics is a minor strategy & medium inter-organizational conflict	.74

Two factors appear to have substantial impact on whether or not Hezbollah engages in transnational attacks. One factor is whether or not there are medium inter-organizational conflicts involving Hezbollah. The most substantial factor is whether or not Hezbollah is engaged in electoral politics as a minor strategy (that is, they have candidates holding elected office but it is not an election year – that would be major strategy). The positive relationship between medium inter-organizational conflict and transnational attacks could reflect a "rally round the flag" phenomenon in which Hezbollah tries to best its local rivals by focusing on the common enemy. But this phenomenon does not appear to apply to major inter-organizational conflicts, possibly because these conflicts cannot be defused as easily and require more attention from the leadership and more resources. This would also explain why medium inter-organizational rioting has a smaller positive effect on the probability of transnational attacks. Rioting requires substantial manpower, leaving fewer resources for the transnational attacks.

The correlation between minor involvement in electoral politics and trans-national attacks highlights the tension between Hezbollah's ideology and practical need for public support. The decision to enter Lebanese politics was a contentious one, with the most strident Hezbollah militants opposed because they feared it would corrupt the organization and distract it from its primary role of confronting Islam's enemies. [10] To placate this faction it is essential that Hezbollah maintain its aggressive stance against Israel, not only by fighting Israeli forces in Lebanon but also by launching attacks into Israel itself. However, Hezbollah usually refrains from these attacks during election years. The exception was 1996, when a Hezbollah rocket campaign provoked a particularly harsh Israeli bombardment in which more than a hundred Lebanese were killed and many more were left homeless. In the elections later that year, Hezbollah lost two seats in Lebanon's parliament. This reflects the tension between Hezbollah's core ideology of confronting Israel and their need not to agitate the many Lebanese who are frustrated at their country's being used as a leading front for the Arab-Israeli conflict.

4 Conclusions

The SOMA system generated rules on Hezbollah's behavior that reveal Hezbollah as a complicated organization that responds to multiple constituencies. Internal conflicts with other Lebanese groups, lack of foreign state political support or support from the diaspora, and efforts to garner this support all impact Hezbollah's operations. Particularly intriguing is the strong positive relationship between participating in electoral politics and attacking transnational targets. This indicates that Hezbollah, despite its participating in an electoral process, remains committed to using violence to further its ends. However, the fact that Hezbollah refrains from these attacks during election years shows that the organization is sensitive to its domestic constituency. It is possible that driving wedges between Hezbollah and its state sponsors, while supporting political parties that vigorously oppose Hezbollah could force the organization to choose between satisfying its Lebanese constituency or its own ideology. Ending this balancing act could marginalize the organization and reduce its capacity for violence.

Since the summer 2006 war, the region has been on tenterhooks awaiting a resumption of hostilities. Examining the SOMA rules reveals some possible reasons why this war has not come. Early in 2007 domestic tensions between different Lebanese parties boiled over into large-scale protests and riots. Hezbollah has a low likelihood of engaging in transnational violence when there are major inter-organizational conflicts. There was also a Presidential election in November 2007. The Lebanese Presidency is reserved for a Maronite Christian, so Hezbollah does not have a candidate; however, it engaged in the process and has a substantial stake in the outcome. With a delicate balance of power in Lebanon, Hezbollah may not wish to inflame its opposition and tilt the political scales at such a sensitive time.

Acknowledgements: The authors gratefully acknowledge funding support for this work provided by the Air Force Office of Scientific Research through the Laboratory for Computational Cultural Dynamics (LCCD) grants AFOSR grants FA95500610405 and FA95500510298, the Department of Homeland Security (DHS) through the National Consortium for the Study of Terrorism and Responses to Terrorism (START) grants N00140510629 and grant N6133906C0149, the Army Research Office under grant DAAD190310202, and the National Science Foundation under grant 0540216. Any opinions, findings or recommendations in this document are those of the authors and do not necessarily reflect the views of sponsors.

References

1. Simari G, Sliva A, Nau D, Subrahmanian V (2006) A stochastic language for modelling opponent agents. In: proceedings International Conference on Autonomous Agents and Multiagent Systems, 244-246.
2. Khuller S, Martinez V. Nau D, Simari G, Sliva A, Subrahmanian V (2007) Action Probabilistic Logic Programs, accepted for publication in *Annals of Mathematics and Artificial Intelligence.*
3. Khuller S, Martinez V, Nau D, Simari G, Sliva A, Subrahmanian V (2007) Finding Most Probable Worlds of Logic Programs. In proceedings Intl. Conf. on Scalable Uncertainty Management, Springer Lecture Notes in Computer Science Vol. 3442, pages 45-59.
4. Sliva A, Martinez V, Simari G, Subrahmanian V (2007) SOMA Modles of the Behaviors of Stakeholders in the Afghan Drug Economy: A Preliminary Report. In: Proceedings of the First International Conference on Computational Cultural Dynamics, pp. 78-86.
5. Subrahmanian V. (2007) Cultural Modeling in Real-Time, *Science*, Vol. 317, Nr. 5844, pages 1509-1510.
6. Armitage R, (2002) America's Challenges in a Changed World. Remarks at the United States Institute of Peace Conference.
7. Mannes A (2004) *Profiles in Terror: The Guide to Middle East Terrorist Organizations.* Rowman & Littlefield, Lanham MD.
8. Asal V, Johnson C, Wilkenfeld J (2007) Ethnopolitical Violence and Terrorism in the Middle East. In Hewitt J, Wilkenfeld J, Gurr T (eds) *Peace and Conflict 2008.* Paradigm, Boulder CO.
9. Minorities at Risk Project. College Park, MD: Center for International Development and Conflict Management. 2005 Retrieved from http://www.cidcm.umd.edu/mar/.
10. Ranstorp M (1998) The Strategy and Tactics of Hizballah's Current 'Lebanonization Process.' Mediterranean Politics, Vol. 3, No. 1, (Summer 1998) pp. 103-134.
11. Schenker D, Exum A (2007) Is Lebanon Headed toward Another Civil War? Washington Institute for Near East Policy *PolicyWatch #1189.*

12. Schenker D (2007) Presidential Elections in Lebanon: Consensus or Conflagration? Washington Institute for Near East Policy *PolicyWatch #1299*.

An approach to modeling group behaviors and beliefs in conflict situations

Norman D. Geddes[†] and Michele L. Atkinson[*]

[†] ngeddes@asinc.com, Applied Systems Intelligence, Inc., Alpharetta, GA
[*] matkinson@asinc.com, Applied Systems Intelligence, Inc. Alpharetta, GA

Abstract In modern theater military operations, increasing attention is being directed to the coordination of military operations with ongoing social and economic redevelopment and the reformation of a viable political process. Recent experiences in Somalia, Bosnia, Afghanistan and Iraq have resulted in the formulation by the US Army of a specific doctrine, known as DIME (Diplomatic, Informational, Military and Economic) to coordinate all aspects of operations. A key technical requirement for DIME is the evolution of models of population belief and behavioral responses to planned tactical operations. In this paper, we describe an approach to modeling the responses of population groups to overt manipulations over time periods of days to weeks. The model is derived from extensive past successful work by the authors in using computational cognitive models to reason about the beliefs, desires and intentions of individuals and groups.

1 Introduction

Current low-intensity theater military operations are multi-dimensional. In Iraq and Afghanistan, and before that, in Somalia and Bosnia, the desired geo-political outcome has involved far more than applying military force to defeat an armed enemy. As a result, the importance of joint efforts in diplomacy, information distribution and economic interventions can be as significant to the outcome as achieving military dominance.

The recognition of the importance of these factors in modern low intensity conflicts has led to the formulation by the US Army of a doctrine known as DIME. DIME (Diplomatic, Information, Military and Economic) requires that a tactical commander consider all of these factors in a coordinated manner as a part of mission planning and execution. The goal of DIME is to avoid counter-productive and conflicting activities in the conduct of tactical operations within a theater. For example, if restoring local agriculture to increase food supplies and encourage economic participation by the population is a high priority economic activity, a tactical operation that would disrupt agriculture and damage crops might be avoided if other more compatible operations could be performed instead.

Unlike traditional aspects of military science, such as weapons effects or logistics, DIME tasks do not yet have a reliable set of models to estimate the range of effects that might result. As a result, the integration of DIME with the overall doctrine of Effects Based Operations is hindered. While increased training to staff officers about DIME and general methods to understand and estimate the effects of DIME tasks may temporarily bridge the gap, the Army needs a reliable, maintainable and deployable solution to estimating the effects of DIME tasks. The solution must be easily configured to the theater and produce understandable and immediately applicable results to be of value to a tactical commander.

Estimating the potential effects of DIME tasks involves modeling populations and the social, economic, informational and ideological forces within the populations (Political, Military, Economic, Social, Informational Infrastructure, PMESII). The model of the population and its PMESII must be predictive with respect to DIME tasks and how these tasks in the current situation will result in changes to the beliefs, desires and intentions of the members of the population. The time frame over which the predictions must be valid is determined by the time frames for preparation and execution of tactical operations. Ideally, model predictions can serve as important indicators of likely progress towards theater goals.

This paper describes an approach to building computational models of short-period beliefs and intentions of groups within populations based on past successful models of both individual and group beliefs and intentions. Section 2 provides a brief review of the importance of groups in social systems as background to the choice of model structure. A brief introduction to some of the previous development of computational cognitive models of beliefs, desires and intentions is given in Section 3. Section 4 describes the main features of the model and Section 5 describes approaches to validation of the model.

2 The role of groups in social systems

While western political thought has been strongly influenced by the assumption that civilization arose from agreements between individuals, this assumption overlooks the role of groups. It is widely accepted by cultural anthropologists that humans formed and maintained cohesive groups long before the emergence of government as we know it today. The existence of family groups and larger tribal aggregations continue to play a dominant role today in emerging political and economic structures within developing countries.

Group behavior has been a subject of considerable study by social scientists since the mid-20th century(Alford, 1994; Brown, 2001; Hogg, 2000). As a result, many aspects of group behavior have been described. Some of the salient findings are:

- While group members influence each other, and may choose varying levels of commitment to the interests of the group, each member thinks and acts as an individual. This principle is known as *methodological individualism*.
- Groups develop different levels of commitment by their members, and the commitment levels are dynamic in nature. Members re-evaluate their role in the group, the demands of the group and the needs of their own individualism on a recurrent basis. There are two relatively stable evolutionary states: one that requires all members to surrender their individuality totally to the interests of the group (de-individuation), and one in which each member has found an acceptable balance between the group and his own individual existence. Between these extremes, groups tend to be unstable.
- Groups interact with other groups to form alliances or rivalries. Groups can be assimilated into other groups, sometimes forcibly. Groups can also dissolve as a result of loss of commitment by its members.
- Leadership plays a pivotal role in the evolution of a group. Leaders who emphasize devotion to the group's ideology will more likely move the group towards de-individuation. The members of the group generally want to be led, thus placing pressure on the leader. Achieving a balanced group requires leadership that emphasizes individual responsibility while still fostering group cohesion.

The large majority of the interactions that a population initiates are decidedly purposeful, rather than random. Individuals, groups and populations act out of biological necessity to feed, reproduce, and protect their survival; out of economic interest to make, trade and consume; out of social interest to communicate and to organize their behaviors to reduce conflicts; and out of religious conviction to worship and in some cases to convert others to their beliefs. Each of these purposeful acts implies a belief in a set of *causal mechanisms* that are sufficiently robust that the outcomes of actions can be reasonably predicted and expected by the actors themselves. It is very important to note that these believed causal mechanisms need not be correct or factual—just believed. It is the existence of the *beliefs* that the causal mechanisms exist that in turn mediates the population actions and reactions.

If we believe that such imputed causal mechanisms are present within a subsystem and are observed to be robust enough to make predictions, then characterizing the causal mechanisms in terms of their inputs, transfer function and outputs provides in principle a means to make predictions about the future changes in the state of a subsystem.

While characterizing the state of a population's belief structures at a particular period of time has been heavily studied and practiced, much less progress has been made in modeling changes in the beliefs, desires and intentions of populations. In the case of causal models of human population subsystems, embedded in their habitats, researchers face a number of current gaps: (1) understanding and representing the underlying causality within the population; (2) formulating models that are both sensitive and computable; and (3) validating the predictions of population beliefs, intentions and behaviors made by the model.

3 Computational cognitive models

Despite the extent of group behavior studies, very few computational models of group behavior and beliefs have been proposed or tested. One reason may be the great difficulty in establishing strong unifying principles for group behavior. The behavior of a group is strongly influenced by the context in which it is situated. As a result, it is difficult to generalize the observed behavior of a specific group from the specific context in which the behavior was observed.

One approach to modeling that has emphasized the role of context is that of cognitive systems. A cognitive model is a model that represents the beliefs, desires and intentions of individuals and groups, and estimates changes in the state of the beliefs, desires and intentions as a result of observations and actions. Cognitive models frequently employ symbolic computing techniques from the field of Artificial Intelligence.

Cognitive models often represent beliefs as a set of symbolic assertions with likelihoods attached. Desires and intentions are often represented as goals and possible methods for achieving the goals. As the model receives observations about its context, it reasons about the relationships between the observations and its current assertions to update its beliefs. Based on its beliefs, the cognitive model may also update its goals and methods. The current believed state may satisfy some goals and give rise to others. Methods that were being performed may be halted, and different methods that match the believed situation may be invoked. Because the goals and methods may persist over time, the current goals and methods provide a prediction of the future behavior of the model.

A particularly useful aspect of cognitive models is their use of knowledge bases as the means to describe the relationships between beliefs, desires and intentions. This allows the knowledge bases to represent the strong role of context, while allowing the software engines within the cognitive model to remain very general. Adaptation of such a model to a new context may require adjustment of the knowledge, but not rewriting the entire model structure.

An approach to modeling group behaviors and beliefs in conflict situations

Applying cognitive models to groups requires careful consideration of the principle of methodological individualism. It is an empirical issue to determine if a group cognitive model can adequately represent the future behavior of members of the group in the aggregate, and thus the group's impact on social and political issues.

One of the earliest uses of a cognitive model to predict the behaviors of ideological groups was Carbonell's software program POLITICS (Carbonell, 1978; 1979). This program generated dialogs from two distinct political viewpoints in response to questions about a national policy issue.

Each of the political viewpoints was represented with its own knowledge base, consisting of a tree of goals. The goal tree was composed of goal dependencies and reached from simple, concrete goals at the bottom of the goal tree to more abstract goals at higher levels. While the goal models in POLITICS allowed good representations of static beliefs, the original implementation did not provide for change in beliefs over time. Another limitation was the lack of any representation of the informational and economic context for goal formation.

Following Carbonell's original work, the authors have pursued the development of cognitive models for group behavior with continued progress (Geddes, 1994; 1997; Geddes and Lizza, 1999; Geddes and Lizza, 2001, Atkinson, 2003; 2004). These models have included both cooperating group beliefs and intentions and hostile group beliefs and intentions. The size of the group settings have ranged from as small as two groups to as large as several hundred groups. The ability of these models to accurately represent dynamic beliefs and intentions has generally exceeded 90% accuracy.

4 Current approach

Our current cognitive model approach has many similarities to the original work by Carbonell, but includes many extensions that are needed to create a dynamic model of beliefs and intentions. These extensions have been made a part of the PreAct® cognitive engine, resulting in a mature and flexible framework for creating cognitive models.

One area of important extensions is the use of the concept graph structure to represent the dynamic state of the surrounding political, economic, social and informational infrastructure. Concept graphs provide a means for distinguishing between the population beliefs about the state of the environment and its true state. Another important feature of concept graphs is the ability to represent uncertain or evidential relationships between dynamic concepts. Concepts are dynamically updated as a result of observations in the form of incoming data about the perceived state of some aspect of the environment.   Figure 1 shows a portion of the Concept Graph for a nation-state that deals with its transportation systems.

Figure 1: A portion of the Concept Graph for a nation-state's PMESII. Observations are collected as reports (lowest level) that are combined within the cognitive model to estimate the believed status of the transportation systems at different levels of aggregation and abstraction.

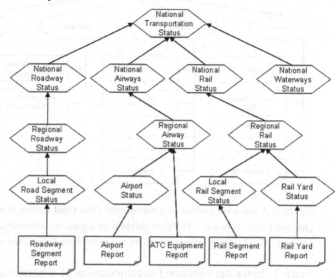

The links between the concepts shown in Figure 1 contain instructions for computing higher level aggregations and abstractions contained in each "parent" concept from the data contained within each "child" concept. Uncertainty calculations are also contained in the links, allowing different sources of information to receive more or less influence in shaping the belief value of the concepts.

A second important extension is the reformulation of Carbonell's original goal tree approach into a Plan-Goal Graph.   This structure allows a principled separation between the desired or intended future state of the environment ("goal") from the means that the goal might be attained ("plan").   In our previous works cited earlier, the Plan-Goal Graph has been used extensively for plan generation and plan recognition in dynamic uncertain environments.   A collection of goals and possible plans for achieving the goals makes up a "course of action" for a group.

As an example, Figure 2 shows the top level Plan Goal Graph for a nation-state. The highest node, **Multi-Interest State,** is a Plan node (denoted by a rectangle) representing the intention of a state to exist and survive.   Sub-goals of the top plan are shown as ovals.   For each goal, several Plan nodes are shown.   The Plan Goal Graph is defined to represent the alternative ways that goals can be achieved, so each plan child of a goal is a possible means to achieve the parent goal. These lower level Plan nodes would also be decomposed into subgoals.   The decomposition in this manner continues until the level of basic interactions is reached.

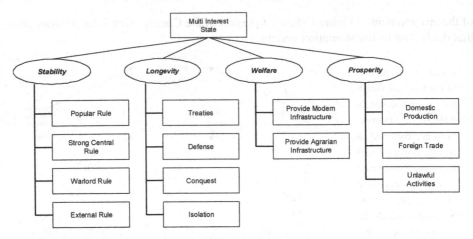

Figure 2: A top-level portion of a nation-state Plan Goal Graph, showing Plans as rectangles and Goals as ovals. The plan children of a goal are alternative means for achieving the goal. Each Plan node can in turn be expanded into its required sub-goals.

Figure 3 shows the continued decomposition of a portion of the **Provide Modern Infrastructure** plan from Figure 2. Only the **Transportation** goal is expanded in the figure. As shown in Figure 3, the **Roadway Built** goal may be achieved by the efforts of either a **Government Group**, a **Business Group** or a **Humanitarian Group**. In this case, these alternatives are not mutually exclusive, allowing for collaboration in achieving the roadways goal. In other cases, group intentions can be in conflict.

Figure 3: A portion of the continued decomposition of the Plan Goal Graph for a nation-state. The Transportation goal may be met by any combination of rail, roadways, and airways. The Roadways plan in turn has subgoals for building and operating the roadways.

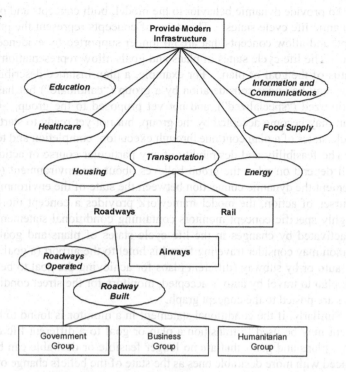

The Plan Goal Graph provides for the intentions of many types of groups within the model. As an example, Figure 4 shows a portion of the graph for a **Legitimate Political Group**. The bottom-level plans of this portion of the graph are at the level of basic interactions.

Figure 4: A portion of the continued decomposition of the Plan Goal Graph for a Legitimate Political Group. The lowest level plan nodes can be observed as a direct result of actions taken by the group members.

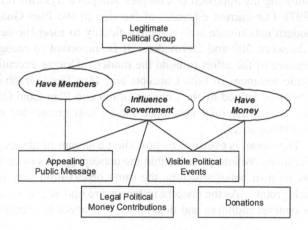

To provide dynamic behavior to the model, both concepts and plans and goals have dynamic life cycle states. The states of concepts represent the prominence of a concept, and allow concepts that are no longer supported by evidence to become forgotten. The life cycle states of plans and goals allow representation of the commitment status of the goal or plan. For example, a plan instance describing a possible behavior may be under consideration by a group ("candidate") but has not yet been fully definitized ("specialized"), and not yet proposed to the group. Once proposed, the plan may become accepted by the group, but not yet ready to start. The allowed life cycle states of a plan continue through execution, completion and termination.

The feasibility and desirability of any particular course of action (plans and goals) will depend on what the group believes about the environment (concepts). To implement the dynamic connection between the state of the environment and the possible courses of action, the model framework provides a concept monitoring mechanism. Highly specific concept monitors containing conditional statements are activated and deactivated by changes in the life cycle states of plans and goals. For example, a person may consider traveling from his hotel to the airport (a goal to be *at the airport*) by auto or by subway (different plans for achieving the goal to be *at the airport*). If the plan to travel by auto is accepted, monitors for the street conditions and traffic delays are posted to the concept graph.

Similarly, if the conditional statement in a monitor is found to be true, the detected event may be used to transition a plan or goal to a different life cycle state. In this way, plans and goals that are no longer feasible or desirable can be discarded and replaced with more desirable ones as the state of the beliefs change over time.

These underlying features of the PreAct® suite allow the development of group-centered models of populations. Each group is represented by a set of concept graph instances, plan and goal instances and monitor instances. The evolution of the population and its infrastructure occurs as a result of the interactions between the groups, following the approach of Complex Adaptive Systems (cf Axelrod, 1997; Cederman, 1997). Our current estimates of the size of the Plan Goal Graph needed to model a modern nation-state with sufficient fidelity to meet the needs of tactical commanders is between 300 and 500 nodes. It is important to recognize that this estimate is a measure of the effort to build the model. During execution of the model, many dynamic instances of both Concepts and Plan Goal Graph nodes are created from the templates defined by the Concept Graph and Plan Goal Graph. Our current expectation is that the fully developed model will create and manage between 5,000 and 10,000 instances.

The model is executed by providing a stream of observations as its input. The observations are interpreted within the concept graphs of each group. Since each group has its own belief structure, the same observation may be interpreted differently by each group. As the concept instances are updated, the model of each group evaluates its concept monitors and determines if it needs to change the life cycle states of the instances of any of its plans and goals. As group plans become active, they are performed by the model, resulting in actions by the group that in turn are fed back into the model as observations to the other groups. In this way, group actions that change the underlying environment or change group relationships can affect the beliefs and intentions of other groups.

An important aspect of the model processing is the association of a temporal sequence of observed actions performed by multiple actors with the correct plan instances for each of the actors. Because observed actions do not always occur in strongly ordered sequences, this association is performed by a non-monotonic abductive reasoner described more completely in Geddes (1989).

5 Short period model performance

While we are presently evaluating the performance of this class of model, it is our conjecture that group-oriented cognitive models have good validity for short periods of time, on the order of days to weeks. Our prior work in Air Traffic Control has shown that, in situations where plans were relatively short (minutes to several hours in duration), the models have been able to predict complex goal-based behavior for several plan lifetimes into the future. We currently anticipate that the plan lifetimes for political and economic activities are at least several days to weeks. As a result, we expect to get good predictive performance for periods of days to several weeks into the future.

We intend to build and evaluate models of this type as a part of our current research agenda. After considering several methods for assessing model accuracy, we have chosen blind historical case analysis as an initial approach. An independent team will select historical cases. Each case will be split by the independent team into a "lead in set" of observations and a "test set" of observations, based on the actual evolution of the selected historical case. The lead in observations and the test set observations will contain the interleaved actions of many interacting groups. The split point in the case will normally be a key decision that is made by one or more of the groups. The model will be run using the lead in set of observations to generate the model's predictions of the expected future behaviors of the groups. The predictions will be compared to the historical case's test set of observations to determine the extent to which the model was successful in predicting the actual evolution of the situation over the succeeding weeks.

A second issue we plan to explore is the generality of the models across cultures. One advantage of the knowledge-based approach used in the cognitive model is that the model's knowledge about cultural differences is expected to be directly adaptable. It is an open question as to the extent of the adaptations needed to sustain a suitable level of accuracy across dissimilar cultures. This aspect can also be tested by using

An approach to modeling group behaviors and beliefs in conflict situations

the blind historical case approach, in which cases from one culture can be used to craft the knowledge and separate cases from a dissimilar culture can be used for testing.

In examining needs for longer term models, we suggest that other modeling techniques may be better suited for predictions over periods of several months to years. Over this longer period of time, specific goals and plans are more difficult to project, and more aggregated models may be more effective.

Acknowledgements. The authors wish to thank the conference reviewers for their helpful suggestions.

References

1. Alford, C.F.. (1994)   *Group Processes and Political Theory.* Yale University Press, New Haven.
2. Atkinson, M. (2003) Contract nets for control of distributed agents in unmanned air vehicles.   In *Proceedings of the 2nd AIAA "Unmanned Unlimited" Conference,* San Diego, CA.
3. Atkinson, M. (2004) Results analysis of using free market auctions to distribute control of unmanned air vehicles. In *Proceedings of the 3rd AIAA "Unmanned Unlimited" Conference.* Chicago. IL.
4. Axelrod, R. (1997) *The Complexity of Cooperation.* Princeton University Press, Princeton, NJ
5. Brown, R. (2001) *Group Processes: Dynamics With and Between Groups.* 2nd Ed. Blackwell Publishing, London.
6. Carbonell, J. (1978) Politics: Automated ideological reasoning, *Cognitive Science,* 2/1: 27-51.
7. Carbonell, J. (1979) *The counterplanning process: a model of decision making in adverse conditions.*   Research Report, Department of Computer Science, Carnegie Mellon University.
8. Cederman, Lars-Erik (1997) *Emergent Actors in World Politics: How states and nations develop and dissolve.*   Princeton University Press, New Jersey.
9. Geddes, N.D. (1989) Understanding human operators' intentions in complex systems. Ph.D. Thesis, Georgia Institute of Technology, Atlanta, GA.
10. Geddes, N.D. (1994) A model for intent interpretation for multiple agents with conflicts. *Proceedings of the IEEE International Conference on Systems, Man and Cybernetics, SMC-94.* San Antonio, TX.
11. Geddes, N.D. (1997) Large scale models of cooperative and hostile intentions, in Proceedings of *IEEE Computer Society International Conference and Workshop on Engineering of Computer Based Systems* (ECBS'97), Monterey, CA.
12. Geddes, N.D. and Lizza, C.S. (1999) Shared plans and situations as a basis for collaborative decision making in air operations. SAE World Aeronautics Conference, Irving CA. SAE Paper 1999-5538.
13. Geddes, N.D. and Lizza, C.S. (2001) Practical Applications of a Real Time, Dynamic Model of Intentions. *AAAI Fall Symposium,* Cape Cod MA.
14. Hogg, M. (2000) *Blackwell Handbook of Social Psychology: Group Processes.* Blackwell Publishing, London.

Computational Models of Multi-National Organizations

A. H. Levis, Smriti K. Kansal, A. E. Olmez, and Ashraf M. AbuSharekh

alevis@gmu.edu, smritipat@aol.com, aolmez@gmu.edu, aabushar@gmu.edu System Architectures Laboratory, George Mason University, Fairfax, VA

Abstract An algorithm for designing multi-national organizations that takes into account cultural dimensions is presented and an example from the command and control field is used to illustrate the approach.

1 Introduction

The problem of modeling multi-national organizations such as those found in military coalition operations has received renewed attention. Coalition partners may have differences in equipment or materiel, differences in command structures, differences in constraints under which they can operate, and, last but not least, differences in culture. The differences in equipment and in operational constraints can be handled easily in the existing modeling framework. Differences in command structures require some additional work to express them in structural and quantitative ways. The real challenge is how to express cultural differences in these, primarily mechanistic, models of organizations.

This work focuses on the ability to introduce attributes that characterize cultural differences into the organization design and use simulation to see whether these parameters result in significant changes in structure. The objective, therefore, is to relate performance to structural features but add attributes that characterize cultural differences. Specifically, the attributes or dimensions defined by Hofstede (2001) are introduced in the design process in the form of constraints on the allowable interactions within the organization. In Section 2, the modeling approach is described briefly since it has been documented extensively in the literature. In Section 3, the Hofstede dimensions are introduced and then applied to the organization design algorithm. In Section 4, an illustrative example is presented, followed by conclusions.

2 The Decision Maker Model and Organizational Design

The five-stage interacting decision maker model (Levis, 1993) had its roots in the investigation of tactical decision making in a distributed environment with efforts to understand cogtive workload, task allocation, and decision-making. This model has been used for fixed as well as variable structure organizations (Perdu and Levis, 1998). The five-stage decision maker (DM) model is shown in Figure 1

The DM receives signals from the external environment or from another decision maker. The Situation Assessment (SA) stage represents the processing of the incoming signal to obtain the assessed situation that may be shared with other DMs. The decision maker can also receive situation assessment signals from other decision makers within the organization; these signals are then fused together in the Information Fusion (IF) stage. The fused information is then processed at the Task Processing (TP) stage to produce a signal that contains the task information necessary to select a response. Command input from superiors is also received. The Command Interpretation (CI) stage then combines internal and external guidance to produce the input to the Response Selection (RS) stage. The RS stage then produces the output to the environment or to other organization members. The key feature of the model is the explicit depiction of the interactions with other organization members and the environment.

Figure 1: Model of the five- stage decision maker

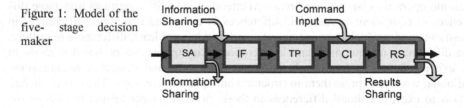

These interactions follow a set of rules designed to avoid deadlock in the information flow. The representation of the interactions can be aggregated into two vectors **e** and **s**, representing interactions with the external environment and four matrices **F**, **G**, **H** and **C** specifying intra-organizational interactions (Fig. 2).

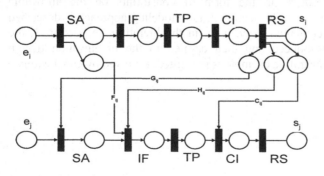

Figure 2: One-sided interactions between DM_i and DM_j

The analytical description of the possible interactions between organization members forms the basis for an algorithm that generates all the architectures that meet some structural constraints as well as application-specific constraints that may be present. The most important constraint addresses the connectivity of the organization - it eliminates information structures that do not represent a single integrated organization Remy and Levis (1988) developed an algorithm, named the Lattice algorithm, that determines the maximal and minimal elements of the set of designs that satisfy all the constraints; the entire set can then be generated from its boundaries. The algorithm is based on the notion of a simple path - a directed path without loops from the source to the sink. Feasible architectures are obtained as unions of simple paths. Consequently, they constitute a partially ordered set. The algorithm receives as input the matrix tuple of dimension n {e, s, F, G, H, C}, where n is the number of organization members.

A set of four different structural constraints is formulated that applies to all organizational structures being considered.

R1   A directed path should exist from the source to every node of the structure and from every node to the sink.

R2   The organizational structures should be acyclical.

R3   There can be at most one link from the RS stage of a DM to each one of the other DMs; i.e., for each i and j, only one element of the triplet $\{G_{ij}, H_{ij}, C_{ij}\}$ can be nonzero.

R4   Information fusion can take place only at the IF and CI stages. Consequently, the SA and RS stages of each DM can have only one input.

To introduce user-defined constraints that will reflect the specific application the organization designer is considering, appropriate 0s and 1s can be placed in the arrays {e, s, F, G, H, C}. The other elements will remain unspecified and will constitute the degrees of freedom of the design.

A feasible structure is one that satisfies both the structural and user-defined constraints. A maximal element of the set of all feasible structures is called a maximally connected organization (MAXO). Similarly, a minimal element is called a minimally connected organization (MINO). The design problem is to determine the set of all feasible structures corresponding to a specific set of constraints. The Lattice algorithm generates, once the set of constraints is specified, the MINOs and the MAXOs that characterize the set of all organizational structures that satisfy the requirements. This methodology provides the designer of organizational structures with a rational way to handle a problem whose combinatorial complexity is very large. Having developed a set of organizational structures that meets the set of logical constraints and is, by construction, free of structural problems, we can now address the problem of incorporating attributes that characterize cultures.

3 Modeling Cultural Attributes

Hofstede (2001) distinguishes dimensions of culture that can be used as an instrument to make comparisons between cultures and to cluster cultures according to behavioural characteristics. Culture is not a characteristic of individuals; it encompasses a number of people who have been conditioned by the same education and life experience. Culture, whether it is based on nationality or group membership such as the military, is what the individual members of a group have in common (De Mooij, 1998). To compare cultures, Hofstede originally differentiated them according to four dimensions: *uncertainty avoidance (UAI), power distance (PDI), masculinity-femininity (MAS),* and *individualism-collectivism (IND).* The dimensions were measured on an index scale from 0 to 100, although some countries may have a score below 0 or above 100 because they were measured after the original scale was defined in the 70's. The hypothesis here is that these dimensions may affect the interconnections between decision makers working together in an organization. Organizations with low power distance values are likely to have decentralized decision making characterized by a flatter organizational structure; personnel at all levels can make decisions when unexpected events occur with no time for additional input from above. In organizations with low scores on uncertainty avoidance, procedures will be less formal and plans will be continually reassessed for needed modifications.

The trade off between time and accuracy can be used to study the affect of both power distance and uncertainty avoidance (Handley and Levis, 2001). Messages exchanged between decision makers can be classified according to three different message types: information, control, and command ones. Information messages include inputs, outputs, and data; control messages are the enabling signals for the initiation of a subtask; and command messages affect the choice of subtask or of response. The messages exchanged between decision makers can be classified according to these different types and each message type can be associated with a subjective parameter. For example, uncertainty avoidance can be associated with control signals that are used to initiate subtasks according to a standard operating procedure. A decision maker with high uncertainty avoidance is likely to follow the procedure regardless of circumstances, while a decision maker with low uncertainty avoidance may be more innovative. Power distance can be associated with command signals. A command center with a high power distance value will respond promptly to a command signal, while in a command center with a low power distance value this signal may not always be acted on or be present.

Cultural constraints help a designer determine classes of similar feasible organizations by setting specific conditions that limit the number of various types of interactions between decision makers. Cultural constraints are represented as interactional constraint statements. An approach for determining the values of these constraints has been developed by Olmez (2006). The constraints are obtained using a linear regression on the four dimensions to determine the change in the range of the number of each type of interaction that is allowed.

$$dY = c + \alpha(PDI) + \beta(UAI) + \gamma(MAS) + \delta(IND)$$

where Y is #F or #G or #H or #C

Example: #F ≤ 2, #G = 0, 1 ≤ #H ≤ 3, #C = 3

C-Lattice Algorithm. This is an extension of the Lattice algorithm that allows cultural constraints to be imposed as additional structural constraints, R5-R8, on the solution space. For the cultural constraint example given above, they become:

- R5: The number of F type interactions must be between 0 and 2
- R6: The number of G type interactions must equal 0
- R7: The number of H type interactions must lie between 1 and 3
- R8: The number of C type interactions must equal 3

Figure 3: Flowchart for culturally constrained solution

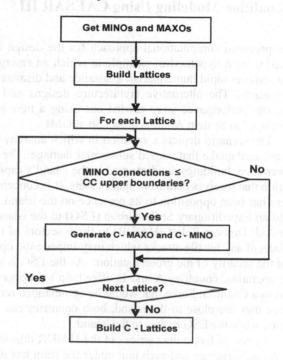

The flowchart in Fig. 3 explains the generation of the culturally constrained solution. MAXOs and MINOs are generated using the same algorithm described in Remy and Levis (1988). The "Build Lattices" step checks if a MINO is contained within a MAXO. If it is, then the MINO is connected to that MAXO and forms part of a lattice. For each lattice, we check the MINO to see if it violates the cultural constraints. For example, if the number of F type interactions in the MINO is two and cultural constraint allows only one, then the MINO does not satisfy the cultural attributes and since the MINO is the minimally connected structure in that lattice, no other structure will satisfy the constraints. Hence the lattice can be discarded. If the MINO does pass the boundary test, then simple paths are added to it to satisfy the cultural constraints R5 to R8. The corresponding minimally connected organization(s) is now called the C-MINO(s) (culturally bound MINO). Similarly, by subtracting simple paths from the MAXO, C-MAXO(s) can be reached. The step "Build C-Lattices" connects the C-MINOs to the C-MAXOs. The advantage of using this approach is that the designer does not have to know the cultural attributes at the start of the analysis. He can add them at a later stage. This also enables him to study the same organization structure under different cultures, which will be useful in our coalition scenario.

4 Coalition Modeling Using CAESAR III

The proposed computational approach for the design of coalition operations is illustrated using a hypothetical example in which an emergency situation in an island nation requires rapid humanitarian assistance and disaster relief as well as securing military assets. The alternative architecture designs and the associated simulations to evaluate performance were carried out using a new application called CAESAR III developed in System Architectures Lab at GMU.

The scenario depicts a situation in which anarchy has risen on an island due to a recent earthquake that caused substantial damage. The infrastructure and many of the government buildings are destroyed in the island's capital. The US maintains a ground station that receives data from space assets. It is concerned about the rising tensions, as there has been opposition to its presence on the island. As a result, the US decides to send an Expeditionary Strike Group (ESG) to the island to provide timely Humanitarian Aid/ Disaster Relief (HA/DR) to three sectors of the island and to counteract the effects of any hostile attacks which may impede the operations of the HA/DR mission and the security of the ground station. As the ESG is away for the first critical day of the operation, countries A and B offer help to support the mission and agree to take part in a Coalition Force that would be commanded remotely by the ESG commander. Since they are close to the island, both countries can deploy elements in a matter of hours, while the ESG rushes to the island.

A team of five units carries out the HA/DR mission. The team is organized in the divisional structure and each unit under the team has its sub-organizations and staff to perform the tasks allocated to it. The five units are: (1) ESGC: Commander; (2) MEUC: Commander of the Marine Expeditionary Unit; (3) ACE: Air Combat Element with its Commander and sub-organizations; (4) GCE: Ground Combat Element with its Commander and sub-organizations; and (5) CSSE: Combat Service Support Element with its Commander and sub-organizations.

It is assumed that country A can provide support as ACE, GCE and CSSE while country B can only provide support as GCE and CSSE. The roles of ESGC and MEUC remain with the US. The countries are able to provide rapid assistance in coordination with each other and the design question becomes the allocation of different tasks to partners in this ad-hoc coalition.

This is a multi-level design problem in which interactions between different decision making units need to be determined both at the higher level (Level-1) as well as at the lower level (Level-2). Level-1 interactions are interactions between culturally homogenous subunits, while the Level-2 problem consists of designing the internal structure of these homogenous subunits on the basis of a defined set of interactional constraints and culture. The structure of the ESG imposes user constraints to design the Level-1 organization. Figure 4 shows the block diagram of this organization as designed in CAESAR III; the matrices describing the interactions are shown below.

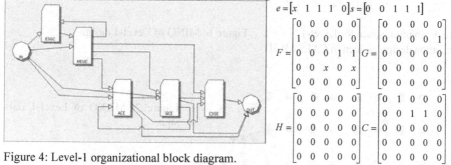

$$e = \begin{bmatrix} x & 1 & 1 & 1 & 0 \end{bmatrix} \quad s = \begin{bmatrix} 0 & 0 & 1 & 1 & 1 \end{bmatrix}$$

$$F = \begin{bmatrix} 0 & 0 & 0 & 0 & 0 \\ 1 & 0 & 0 & 0 & 0 \\ 0 & 0 & 0 & 1 & 1 \\ 0 & 0 & x & 0 & x \\ 0 & 0 & 0 & 0 & 0 \end{bmatrix} \quad G = \begin{bmatrix} 0 & 0 & 0 & 0 & 0 \\ 0 & 0 & 0 & 0 & 1 \\ 0 & 0 & 0 & 0 & 0 \\ 0 & 0 & 0 & 0 & 0 \\ 0 & 0 & 0 & 0 & 0 \end{bmatrix}$$

$$H = \begin{bmatrix} 0 & 0 & 0 & 0 & 0 \\ 0 & 0 & 0 & 0 & 0 \\ 0 & 0 & 0 & 0 & 0 \\ 0 & 0 & 0 & 0 & 0 \\ 0 & 0 & 0 & 0 & 0 \end{bmatrix} \quad C = \begin{bmatrix} 0 & 1 & 0 & 0 & 0 \\ 0 & 0 & 1 & 1 & 0 \\ 0 & 0 & 0 & 0 & 0 \\ 0 & 0 & 0 & 0 & 0 \\ 0 & 0 & 0 & 0 & 0 \end{bmatrix}$$

Figure 4: Level-1 organizational block diagram.

Figure 5 shows the result of running the lattice algorithm on level-1 organization. The solution space contains one MINO, Fig. 6, and one MAXO, Fig. 7. The designer can pick a structure from this space and use it to design the sub-organizations at level-2.

Level-1 design is free of cultural constraints. However Level-2 design uses the C-Lattice algorithm to include cultural attributes to form the various coalition options. The sub-organizations of ACE, GCE and CSSE are designed using CAESAR III. Figures 8, 9 and 10 show the respective block diagrams along with the matrices specifying the user constraints. Since the US always performs the roles of ESGC and MEUC, these sub-organizations are not decomposed further.

Table 1 gives the Hofstede's scores for US, Country A and Country B. Using a multiple linear regression model, these scores are converted into limits to be placed on allowable interactions based on culture. These are imposed as additional structural constraints on the solution space of the sub-organizations. The cultural constraints for the three sub-organizations are shown in tables 2, 3 and 4.

Maximum indicates the limit placed on the number of interactions by user constraints.

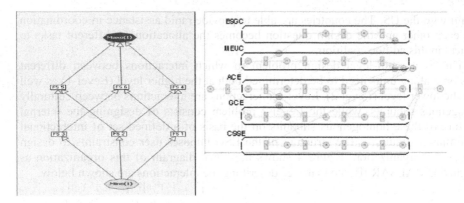

Figure 5: Solution space for Level-1 organization design as seen in CAESAR III

Figure 6: MINO of Level-1 design

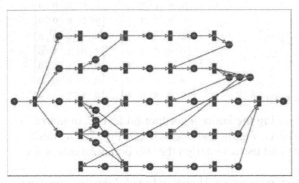

Figure 7: MAXO of Level-1 design

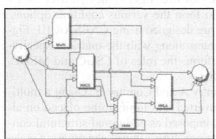

Figure 8: Block diagram for ACE

$$e = \begin{bmatrix} x & 1 & 1 & 0 \end{bmatrix} \quad s = \begin{bmatrix} 0 & x & 1 & 1 \end{bmatrix}$$

$$F = \begin{bmatrix} 0 & 0 & 0 & 0 \\ x & 0 & 0 & x \\ 0 & x & 0 & x \\ 0 & 0 & 0 & 0 \end{bmatrix} \quad G = \begin{bmatrix} 0 & 0 & 0 & 0 \\ 0 & 0 & 0 & 0 \\ 0 & 0 & 0 & 0 \\ 0 & 0 & 0 & 0 \end{bmatrix}$$

$$H = \begin{bmatrix} 0 & 0 & 0 & 0 \\ 0 & 0 & 0 & 0 \\ 0 & x & 0 & x \\ 0 & 0 & x & 0 \end{bmatrix} \quad C = \begin{bmatrix} 0 & 1 & x & x \\ 0 & 0 & 1 & x \\ 0 & 0 & 0 & 0 \\ 0 & 0 & 0 & 0 \end{bmatrix}$$

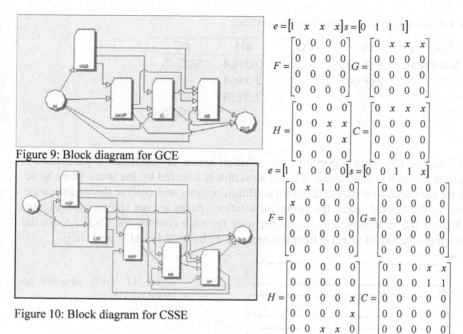

Figure 9: Block diagram for GCE

$$e = \begin{bmatrix} 1 & x & x & x \end{bmatrix} s = \begin{bmatrix} 0 & 1 & 1 & 1 \end{bmatrix}$$

$$F = \begin{bmatrix} 0 & 0 & 0 & 0 \\ 0 & 0 & 0 & 0 \\ 0 & 0 & 0 & 0 \\ 0 & 0 & 0 & 0 \end{bmatrix} G = \begin{bmatrix} 0 & x & x & x \\ 0 & 0 & 0 & 0 \\ 0 & 0 & 0 & 0 \\ 0 & 0 & 0 & 0 \end{bmatrix}$$

$$H = \begin{bmatrix} 0 & 0 & 0 & 0 \\ 0 & 0 & x & x \\ 0 & 0 & 0 & x \\ 0 & 0 & 0 & 0 \end{bmatrix} C = \begin{bmatrix} 0 & x & x & x \\ 0 & 0 & 0 & 0 \\ 0 & 0 & 0 & 0 \\ 0 & 0 & 0 & 0 \end{bmatrix}$$

Figure 10: Block diagram for CSSE

$$e = \begin{bmatrix} 1 & 1 & 0 & 0 & 0 \end{bmatrix} s = \begin{bmatrix} 0 & 0 & 1 & 1 & x \end{bmatrix}$$

$$F = \begin{bmatrix} 0 & x & 1 & 0 & 0 \\ x & 0 & 0 & 0 & 0 \\ 0 & 0 & 0 & 0 & 0 \\ 0 & 0 & 0 & 0 & 0 \\ 0 & 0 & 0 & 0 & 0 \end{bmatrix} G = \begin{bmatrix} 0 & 0 & 0 & 0 & 0 \\ 0 & 0 & 0 & 0 & 0 \\ 0 & 0 & 0 & 0 & 0 \\ 0 & 0 & 0 & 0 & 0 \\ 0 & 0 & 0 & 0 & 0 \end{bmatrix}$$

$$H = \begin{bmatrix} 0 & 0 & 0 & 0 & 0 \\ 0 & 0 & 0 & 0 & 0 \\ 0 & 0 & 0 & 0 & x \\ 0 & 0 & 0 & 0 & x \\ 0 & 0 & x & x & 0 \end{bmatrix} C = \begin{bmatrix} 0 & 1 & 0 & x & x \\ 0 & 0 & 0 & 1 & 1 \\ 0 & 0 & 0 & 0 & 0 \\ 0 & 0 & 0 & 0 & 0 \\ 0 & 0 & 0 & 0 & 0 \end{bmatrix}$$

Table 1. Hofstede's scores for the three countries

Country	PDI	IND	MAS	UAI
US	40	91	62	46
A	38	80	14	53
B	66	37	45	85

Table 2. Cultural Constraints corresponding to ACE

Country	#F	#G	#H	#C
Maximum	0≤F≤4	0	0≤H≤3	2≤C≤5
US	3≤F≤4	0	2≤H≤3	3
A	2	0	2≤H≤3	3
B	2	0	1	4≤C≤5

Table 3. Cultural Constraints corresponding to GCE

Country	#F	#G	#H	#C
Maximum	0	0≤G≤3	0≤H≤3	0≤C≤3
US	0	2	2≤H≤3	2
A	0	2	2≤H≤3	1
B	0	2≤G≤3	2	2≤C≤3

Table 4. Cultural Constraints corresponding to CSSE

Country	#F	#G	#H	#C
Maximum	1≤F≤3	~	0≤H≤4	3≤C≤5
US	2≤F≤4	~	3≤H≤4	□
□	□	~	3≤H≤4	□
□	□	~	□	4≤C≤5

Using the C-Lattice algorithm, the solution space for each sub-organization is computed for each culture and a suitable structure is selected by the user. These structures are then used to form the different coalition options and analyse the performance. In view of the limited space, the complete solution spaces are not shown here. Figures 11-13 show the structures selected by the user for each country for CSSE. A similar approach can be use to select different structures to be used for ACE and GCE.

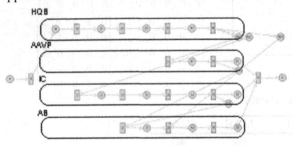

Figure 11: GCE structure selected for US

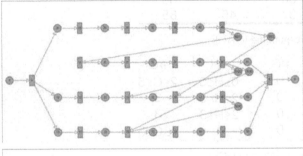

Figure 12: GCE structure selected for Country A

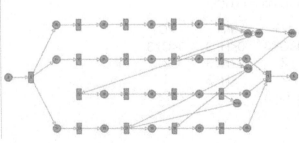

Figure 13: GCE structure selected for Country B

Once the structure is selected, CAESAR III exports it as a Colored Petri net to *CPN Tools* where it can be simulated to analyse performance. For the given scenario, based on the availability of support from the two countries, eight coalition options are possible, excluding the homogeneous option of all US. The five sub-organizations are combined together using Level-1 MINO and the eight options were simulated to study performance in terms of tasks served. The following assumptions were made. Each process (transition) needs 50 units of processing time. Each additional incoming link increases this time by 50 units. The reasoning is that the additional input(s) will require more processing. Hence, structures that have more interactions will take more time to process the tasks, which will affect the overall performance. Figure 14 shows the results of this analysis for all combinations. The x-axis shows the percentage of tasks **un-served**.

Based on these results, the US-US-US-B-A coalition structure performs best. Most options with country B in the CSSE role perform badly. This is because country B needs a high number of command relationships and the structure of CSSE allows for this to occur, thereby increasing the processing delay. User constraints on GCE allow for very similar cultural constraints for all countries; changing the ordering in this role does not change the performance very much. Similar results were obtained when the coalition options were simulated using a Level-1 MAXO organization.

Figure 14: Percent of tasks un-served for coalition options.

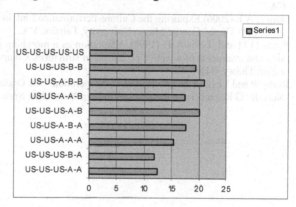

5 Conclusion

A previously developed methodology for the computational design of information processing and decision making organizations has been enhanced to include cultural constraints that affect the choice of organizational structures. While the Hofstede cultural dimensions have been used, other cultural metrics can be used to derive the cultural constrains. A simple example illustrates the approach for designing coalition organizations and analysing their performance. The results indicate that culture does affect the structure and working of organizations thereby affecting the overall performance. This could aid in the allocation of different tasks to partners in an ad-hoc coalition.

Acknowledgements: This work is based on work sponsored by the Air Force office of Scientific Research under agreement number FA9550-05-1-0388. The U.S. Government is authorized to reproduce and distribute reprints for governmental purposes notwithstanding any copyright notation thereon. The views and conclusions contained herein are those of the authors and should not be interpreted as representing the official policies or endorsements, either expressed or implied, of the Air Force Research Laboratory or the U.S. Government.

References

1. Handley H, Levis, A H (2001) Incorporating heterogeneity in command center interactions. Information Knowledge Systems Management 2(4).
2. Hofstede G (2001) Culture's Consequences: Comparing Values, Behaviors, Institutions, and Organizations Across Nations, 2nd edn. Sage, CA .
3. Levis A H (1993). A Colored Petri Net Model of Command and Control Nodes. In: Jones R (ed.) Toward a Science of Command Control and Communications. AIAA Press, Washington, DC.
4. De Mooij M (1998) Global Marketing and Advertising: Understanding Cultural Paradoxes. Sage, CA
5. Olmez, A E (2006) Exploring the Culture-Performance Link in Heterogeneous Coalition Organizations. PhD Thesis, George Mason University, Fairfax, VA.
6. Perdu D M and Levis A H (1998) Adaptation as a morphing process: A methodology for the design and evaluation of adaptive organization structures. Computational and Mathematical Organization Theory 4(1): 5-41
7. Remy P and Levis A H (1988) On the Generation Of Organizational Architectures Using Petri Nets. In: G Rozenberg (ed) Advances in Petri Nets 1988, Springer-Verlag, Berlin.

Clustering of Trajectory Data obtained from Soccer Game Records – A First Step to Behavioral Modeling –

Shoji Hirano[1] and Shusaku Tsumoto[2]

[1]hirano@med.shimane-u.ac.jp, Shimane University, Izumo, Japan
[2]tsumoto@med.shimane-u.ac.jp, Shimane University, Izumo, Japan

Abstract Ball movement in a soccer game can be measured as a trajectory on two dimentional plane, which summarizes the tactic or strategy of game players. This paper gives a first step to extract knowledge about strategy of soccer game by using clustering of trajectory data, which consists of the following two steps. First, we apply a pairwise comparison of two trajectories using multiscale matching. Next, we apply rough-set based clustering technique to the similarity matrix obtained by the pairwise comparisons. Experimental results demonstrated that the method could discover some interesting pass patterns that may be associated with successful goals.

1 Introduction

Human movements in a limited space may have similar characteristics if their targets are the same as others. For example, if a person who wants to buy some food with alcoholic drinks go to a supermarket, he/she may buy both drinks and food which will be matched to the drinks. Conventional data mining methods can capture such a co-occurence between two items. However, if we take a close look at the nature of human movements in the supermarket, we may find more precise characteristics of the clients, such as chronological nature of their behavior, which may show that a royal customer tends to go to fish market before they go to buy beer.

One of the most simple measurements of human spatio-temporal behavior is to set up the two-dimensional coordinates in the space and to measure each movement as a trajectory. We can assume that each trajectory includes information about the preference of a customer. Since the collection of trajectories can be seen as a spacio-temporal sequence in the plane, we can extract the common nature of sequences after we apply clustering method to them. When we go back to the data with the clusters obtained, we may have a visualization of common spatio-temporal behavior from the data, which will be the first step to behavioral modeling.

In this paper, we propose a novel method for visualizing interesting patterns hidden in the spatio-temporal data. As a tangible data we employ soccer game records, as they involve the most important problems in spatio-temporal data mining – the temporal irregularity of data points. Especially, we focus on discovering the features of pass transactions, which resulted in successful goals, and representing the difference of strategies of a team by the pass strategies.

There are two points that should be technically solved. First, the length of a sequence, number of data points constituting a sequence, and intervals between data points in a sequence are all irregular. A pass sequence is formed by concatenating contiguous pass events; since the distance of each pass, the number of players translating the contiguous passes are by nature difference, the data should be treated as irregular sampled time-series data. Second, multiscale observation and comparison of pass sequences are required. This is because a pass sequence represents both global and local strategies of a team. For example, as a global strategy, a team may frequently use side-attacks than counter-attacks. As a local strategy, the team may frequently use one-two pass. Both levels of strategies can be found even in one pass sequence; one can naturally recognize it from the fact that a video camera does zoom-up and zoom-out of a game scene. In order to solve these problems, we employed multiscale matching [1], [2], a pattern recognition based contour comparison method. And we employed rough clustering [3], which are suitable of handing relative dissimilarity produced by multiscale matching.

The rest of this paper is organized as follows. Section 2 describes the data structure and preprocessing. Section 3 describes multiscale matching. Section 4 describes rough clustering. Section 5 shows experimental results on the FIFA world cup 2002 data and Section 6 concludes the results. This paper presents a novel method for visualizing interesting patterns hidden in the spatio-temporal data. As a tangible data we employ soccer game records, as they involve the most important problems in spatio-temporal data mining – the temporal irregularity of data points. Especially, we focus on discovering the features of pass transactions, which resulted in successful goals, and representing the difference of strategies of a team by the pass strategies.

The rest of this paper is organized as follows. Section 2 describes the data structure and preprocessing. Section 3 describes multiscale matching. Section 4 describes rough clustering. Section 5 shows experimental results on the FIFA world cup 2002 data and Section 6 concludes the results.

2 Data Structure and Preprocessing

2.1 Data structure

We used the high-quality, value-added commercial game records of soccer games provided by Data Stadium Inc., Japan. The current states of pattern recognition technique may enable us to automatically recognize the positions of ball and players [4], [5], [6], however, we did not use automatic scene analysis techniques because it is still hard to correctly recognize each action of the players.

The data consisted of the records of all 64 games of the FIFA world cup 2002, including both heats and finals, held during May-June, 2002. For each action in a game, the following information was recorded: time, location, names(number) of the player, the type of event (pass, trap, shoot etc.), etc. All the information was generated from the real-time manual interpretation of video images by a well-trained soccer player, and manually stored in the database. Table 1 shows an example of the data. In Table 1, 'Ser' denotes the series number, where a series denotes a set of con-

Table 1 An example of the soccer data record

Ser	Time	Action	T_1	P_1	T_2	P_2	X_1	Y_1	X_2	Y_2
1	20:28:12	KICK OFF	Senegal	10			0	-33		
1	20:28:12	PASS	Senegal	10	Senegal	19	0	-50	-175	50
1	20:28:12	TRAP	Senegal	19			-175	50		
1	20:28:12	PASS	Senegal	19	Senegal	14	-122	117	3004	451
1	20:28:14	TRAP	Senegal	14			3004	451		
⋮					⋮					
169	22:18:42	P END	France	15			1440	-685		

tiguous events marked manually by expert. The remaining fields respectively represent the time of event occurrence ('Time'), the type of event ('Action'), the team ID ('T_1') and player ID ('P_1') of one acting player 1, the team ID ('T_2') and player ID ('P_2') of another acting player 2, spatial position of player 1 ('X_1', 'Y_1'), and spatial position of player 2 ('X_1', 'Y_1'), Player 1 represents the player who mainly performed the action. As for pass action, player 1 represents the sender of a pass, and player 2 represents the receiver of the pass. Axis X corresponds to the long side of the soccer field, and axis Y corresponds to the short side. The origin is the center of the soccer field. For example, the second line in Table 1 can be interpreted as: Player no. 10 of Senegal, locating at (0,-50), sent a pass to Player 19, locating at (-175,50).

2.2 Target Series Selection

We selected the series that contains important PASS actions that resulted in goals as follows.

1. Select a series containing an IN GOAL action.
2. Select a contiguous PASS event. In order not to divide the sequence into too many subsequences, we regarded some other events as contiguous events to the PASS event; for example, TRAP, DRIBBLE, CENTERING, CLEAR, BLOCK. Intercept is represented as a PASS event in which the sender's team and receiver's team are different. However, we included an intercept into the contiguous PASS events for simplicity.

Fig. 1 Spatial representation of a PASS sequences. The vertical line represents the axis connecting the goals of both teams. Near the upper end (+5500) is the goal of France, and near the lower end is the goal of Senegal.

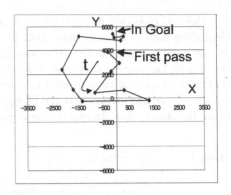

Table 2 Raw data corresponding the sequence in Figure 1.

Ser	Time	Action	T_1	P_1	T_2	P_2	X_1	Y_1	X_2	Y_2
47	20:57:07	PASS	France	16	France	18	-333	3877	122	-2958
47	20:57:08	PASS	France	18	France	17	122	2958	-210	-2223
47	20:57:10	DRIBBLE	France	17			-210	2223	-843	-434
47	20:57:14	PASS	France	17	France	4	-843	434	298	-685
47	20:57:16	PASS	France	4	France	6	298	685	1300	217
47	20:57:17	TRAP	France	6			1300	217		
47	20:57:19	CUT	Senegal	6			-1352	-267		
47	20:57:19	TRAP	Senegal	6			-1352	-267		
47	20:57:20	PASS	Senegal	6	Senegal	11	-1704	702	-2143	2390
47	20:57:21	DRIBBLE	Senegal	11			-2143	2390	-1475	5164
47	20:57:26	CENTERING	Senegal	11			-1475	5164		
47	20:57:27	CLEAR	France	17			175	4830		
47	20:57:27	BLOCK	France	16			281	5181		
47	20:57:27	CLEAR	France	16			281	5181		
47	20:57:28	SHOT	Senegal	19			-87	5081		
47	20:57:28	IN GOAL	Senegal	19			-140	5365		

3. From the Selected contiguous PASS event, we extract the locations of Player 1, X_1 and Y_1, and make a time series of locations $p(t) = \{(X_1(t), Y_1(t)) | 1 \leq t \leq T\}$ by concatenating them. For simplicity, we denote $X_1(t)$ and $Y_1(t)$ by x(t) and y(t) respectively.

Figure 1 shows an example of spatial representation of a PASS sequence generated by the above process. Table 2 provides an additional information, the raw data that correspond to Figure 1. Figure 1 show the PASS sequences representing the following scene: Player no. 16 of France, locating at (-333,3877), send a pass to player 18. Senegal cuts the pass at near the center of the field, and started attack from the left side. Finally, Player no. 11 of Senegal made a CENTERING, and after several block actions of France, Player no. 19 of Senegal made a goal.

By applying the above preprocess to all the IN GOAL series, we obtained N sequences of passes $P = \{p_i | 1 \leq i \leq N\}$ that correspond to N goals, where i of p_i denote the i-th goal.

2.3 Data Cleansing and Interpretation

Continuous actions occurred at the same location should be considered as a single action. For example, in Table 2, the 7th and 8th actions consisting of CUT and TRAP should be treated as a single action, because their interaction does not actually involve any movement of a ball on the field. The 13rd and 14th actions consisting of BLOCK and CLEAR should be similarly treated as a single action. In such a case, we employed only the first action and removed other redundant actions.

We here do not use the time information provided in the data for each action, because the time resolution is insufficient for calculating the moving speed of a ball. Instead, with a fixed interval we performed re-sampling of a trajectory of ball between two successive actions. In this experiment we linearly interpolated the location data at every 55 locational unit (Field length / 200).

3 Multiscale Comparison of Pass Sequences

For every pair of PASS sequences $\{(p_i, p_j) \in P | 1 \leq i < N, i < j \leq N\}$, we apply multiscale matching to compare their dissimilarity. Based on the resultant dissimilarity matrix, we perform grouping of the sequences using rough clustering [3].

Multiscale Matching is a method to compare two planar curves by partly changing observation scales. We here briefly explain the basic of multiscale matching. Details of matching procedure are available in [2].

Let us denote two input sequences to be compared, p_i and p_j, by A and B. First, let us consider a sequence $x(t)$ containing X_1 values of A. Multiscale representation of $x(t)$ at scale σ, $X(t, \sigma)$ can be obtained as a convolution of $x(t)$ and a Gaussian function with scale factor σ as follows.

$$X(t, \sigma) = \int_{-\infty}^{+\infty} x(u) \frac{1}{\sigma\sqrt{2\pi}} e^{-(t-u)^2/2\sigma^2} du \qquad (1)$$

where the gauss function represents the distribution of weights for adding the neighbors. It is obvious that a small σ means high weights for close neighbors, while a large σ means rather flat weights for both close and far neighbors. A sequence will become more flat as σ increases, namely, the number of inflection points decreases. Multiscale representation of $y(t)$, $Y(t, \sigma)$ is obtained similarly. The m-th order derivative of $X(t, \sigma)$, $X^{(m)}(t, \sigma)$, is derived as follows.

$$X^{(m)}(t, \sigma) = \frac{\partial^m X(t, \sigma)}{\partial t^m} = x(t) \otimes g^{(m)}(t, \sigma). \qquad (2)$$

According to the Lindeberg's notions [7], it is preferable to use the modified Bessel function instead of Gaussian function as a convolution kernel for discrete signals. Below we formalize the necessary functions:

$$X(t,\sigma) = \sum_{n=-\infty}^{\infty} e^{-\sigma}I_n(\sigma)x(t-n) \qquad (3)$$

where $I_n(\sigma)$ denotes the modified Bessel function of n-th order. The first- and second-order derivatives of $X(t,\sigma)$ are given as follows.

$$X'(t,\sigma) = \sum_{n=-\infty}^{\infty} -\frac{n}{\sigma}e^{-\sigma}I_n(\sigma)x(t-n) \qquad (4)$$

$$X''(t,\sigma) = \sum_{n=-\infty}^{\infty} \frac{1}{\sigma}(\frac{n^2}{\sigma} - 1)e^{-\sigma}I_n(\sigma)x(t-n) \qquad (5)$$

The curvature of point t at scale σ is obtained as follows.

$$K(t,\sigma) = \frac{X'Y'' - X''Y'}{(X'^2 + Y'^2)^{3/2}}, \qquad (6)$$

where X', X'', Y' and Y'' denote the first- and second-order derivatives of $X(t,\sigma)$ and $Y(t,\sigma)$ by t, respectively.

Next, we divide the sequence $K(t,\sigma)$ into a set of convex/concave subsequences called segments based on the place of inflection points. A segment is a subsequence whose ends correspond to the two adjacent inflection points, and can be regarded as a unit representing substructure of a sequence.

Let us assume that a pass sequence $A^{(k)}$ at scale k is composed of R segments. Then $A^{(k)}$ is represented by

$$A^{(k)} = \left\{ a_i^{(k)} \mid i = 1, 2, \cdots, R^{(k)} \right\}, \qquad (7)$$

where $a_i^{(k)}$ denotes the i-th segment of $A^{(k)}$ at scale $\sigma^{(k)}$. By applying the same process to another input sequence B, we obtain the segment-based representation of B as follows.

$$B^{(h)} = \left\{ b_j^{(h)} \mid j = 1, 2, \cdots, S^{(h)} \right\} \qquad (8)$$

where $\sigma^{(h)}$ denote the observation scale of B and $S^{(h)}$ denote the number of segments at scale $\sigma^{(h)}$.

After that, we trace the hierarchy of inflection points from the top scale to bottom scale based on the proximity of inflection points. This trace is important to capture the hierarchy of segment replacement and to guarantee the connectivity of segments represented at at different scales.

The main procedure of multiscale matching is to find the best set of segment pairs that minimizes the total segment difference. The search is performed throughout all the scales. Figure 2 illustrates the process. For example, three contiguous segments $B_3^{(0)}$, $B_4^{(0)}$ and $B_5^{(0)}$ of sequence B at scale 0 have no similar segments of sequence A at scale 0. However, at a global scale, they can be represented (merged) as a single segment $B_1^{(2)}$ at scale 2, whose shape is similar to $A_1^{(1)}$ or $A_1^{(2)}$ of sequence A at scales

Fig. 2 An illustrative example of multiscale description and matching.

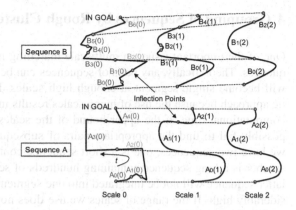

1 or 2. As their origin is $A_3^{(0)}$, we can conclude that the set of segments $B_3^{(0)}$, $B_4^{(0)}$ and $B_5^{(0)}$ are structurally similar to segment $A_3^{(0)}$. On the contrary, segments such as $B_0^{(0)}$ and $B_1^{(0)}$ have locally similar segments $A_0^{(0)}$ and $A_1^{(0)}$ respectively. In this way, we can compare the structural similarity of sequences by changing the observation scales.

There are two restrictions in determining the best set of the segments. First, the resultant set of the matched segment pairs must not be redundant or insufficient to represent the original sequences. Namely, by concatenating all the segments in the set, the original sequence must be completely reconstructed without any partial intervals or overlaps.

Second, the segment dissimilarities accumulated over all matched pairs must be minimized. Dissimilarity $d(a_i^{(k)}, b_j^{(h)})$ of two segments $a_i^{(k)}$ and $b_i^{(h)}$ is defined as follows.

$$d(a_i^{(k)}, b_j^{(h)}) = \frac{\mid \theta_{a_i}^{(k)} - \theta_{b_j}^{(h)} \mid}{\theta_{a_i}^{(k)} + \theta_{b_j}^{(h)}} \left| \frac{l_{a_i}^{(k)}}{L_A^{(k)}} - \frac{l_{b_j}^{(h)}}{L_B^{(h)}} \right| \tag{9}$$

where $\theta_{a_i}^{(k)}$ and $\theta_{b_j}^{(h)}$ denote rotation angles of tangent vectors along segments $a_i^{(k)}$ and $b_j^{(h)}$, $l_{a_i}^{(k)}$ and $l_{b_j}^{(h)}$ denote the length of segments, $L_A^{(k)}$ and $L_B^{(h)}$ denote the total length of sequences A and B at scales $\sigma^{(k)}$ and $\sigma^{(h)}$, respectively.

The total difference between sequences A and B is defied as a sum of the dissimilarities of all the matched segment pairs as

$$D(A, B) = \sum_{p=1}^{P} d(a_p^{(0)}, b_p^{(0)}), \tag{10}$$

where P denotes the number of matched segment pairs. The matching process can be fasten by implementing dynamic programming scheme [2].

4 Grouping of Sequences by Rough Clustering

One of the important issues in multiscale matching is treatment of 'no-match' sequences. Theoretically, any pairs of sequences can be matched because a sequence will become single segment at enough high scales. However, this is not a realistic approach because the use of many scales results in the unacceptable increase of computational time. If the upper bound of the scales is too low, the method may possibly fail to find the appropriate pairs of subsequences. For example, suppose we have two sequences, one is a short sequence containing only one segment and another is a long sequence containing hundreds of segments. The segments of the latter sequence will not be integrated into one segment until the scale becomes considerably high. If the range of scales we use does not cover such a high scale, the two sequences will never be matched. In this case, the method should return infinite dissimilarity, or a special number that identifies the failed matching. This property prevents conventional agglomerative hierarchical clusterings (AHCs) [8] from working correctly. Complete-linkage (CL-) AHC will never merge two clusters if any pair of 'no-match' sequences exist between them. Average-linkage (AL-) AHC fails to calculate average dissimilarity between two clusters.

In order to handle the 'no-match' problem, we employed rough clustering [3], that can handle relatively defined dissimilarity. This method is based on iterative refinement of N binary classifications, where N denotes the number of objects. First, an equivalence relation, that classifies all the other objects into two classes, is assigned to each of N objects by referring to the relative proximity. Next, for each pair of objects, the number of binary classifications in which the pair is included in the same class is counted. This number is termed the indiscernibility degree. If the indiscernibility degree of a pair is larger than a user-defined threshold value, the equivalence relations may be modified so that all of the equivalence relations commonly classify the pair into the same class. This process is repeated until class assignment becomes stable. Consequently, we may obtain the clustering result that follows a given level of granularity, without using geometric measures.

5 Experimental Results

We applied the proposed method to the action records of 64 games in the FIFA world cup 2002 described in Section 2. First let us summarize the procedure of experiments.

1. Select all IN GOAL series from original data.
2. For each IN GOAL series, generate a time-series sequence containing contiguous PASS events. In our data, there was in total 168 IN GOAL series excluding own goals. Therefore, we had 168 time-series sequences, each of which contains the sequence of spatial positions $(x(t), y(t))$.

3. For each pair of the 168 sequences, compute dissimilarity of the sequence pair by multiscale matching. Then construct a 168×168 dissimilarity matrix.
4. Perform cluster analysis using the induced dissimilarity matrix and rough clustering method.

The following parameters were used in multiscale matching: the number of scales = 100, scale interval = 0.5, start scale = 0.5, cost weight for segment replacement = 20.0. We used the following parameters for rough clustering: $\sigma = 2.0$, $T_h = 0.3$. These parameters were determined through preparatory experiments.

Out of 14,196 comparisons, 7,839 (55.2%) resulted in 'matching failure' for which we assigned a special value of '-1' as their dissimilarity. For this highly disturbed dissimilarity matrix, rough clustering produced a total of 12 clusters, each of which contains 4, 87, 27, 17, ... sequences respectively. Figures 3 - 6 respectively show examples of sequences grouped into the four major clusters: cluster 2 (87 cases), 3 (24 cases), 4 (17 cases), 6 (16 cases). Cluster 2 contained remarkably short sequences. They represented special events such as free kicks, penalty kicks and corner kicks, that made goals after one or a few touches. On the contrary, cluster 3 contained complex sequences, each of which contained many segments and often included loops. These sequences represented that the goals were succeeded after long, many steps of pass actions, including some changes of the ball-owner team. Clusters 4 and 6 contained rather simple sequences, most of which contained only several segments. These sequences represented that the goals were obtained after interaction of a few players. These observations demonstrate that the sequences were clustered according to the structural complexity of the pass routes.

6 Conclusions

In this paper, we have presented a new method for cluster analysis of spatio-temporal data with an application to finding interesting pass patterns from time-series soccer game records. Taking two characteristics of the pass sequence – irregularity of data and requirements of multiscale observation – into account, we developed a cluster analysis method based on multiscale matching and rough clustering, which may build a new scheme of sports data mining. Although the experiments are in the preliminary stage and subject to further quantitative evaluation, the proposed method demonstrated its potential for finding interesting patterns in real soccer data. The future work include the use of ball speed, use of other feature points than inflection points, and optimization of segment difference parameters.

Acknowledgements The authors would like to express their sincere appreciation to Data Stadium Inc. for providing the valuable dataset.

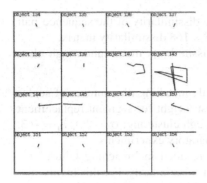

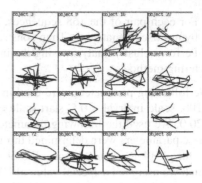

Fig. 3 Sequences in cluster 2 (87 cases).

Fig. 4 Sequences in cluster 3 (24 cases).

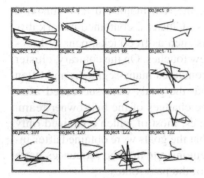

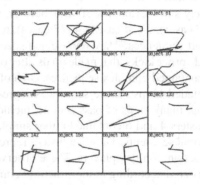

Fig. 5 Sequences in cluster 4 (17 cases).

Fig. 6 Sequences in cluster 6 (16 cases).

References

1. F. Mokhtarian and A. K. Mackworth (1986): Scale-based Description and Recognition of planar Curves and Two Dimensional Shapes. IEEE Transactions on Pattern Analysis and Machine Intelligence, PAMI-8(1): 24-43
2. N. Ueda and S. Suzuki (1990): A Matching Algorithm of Deformed Planar Curves Using Multiscale Convex/Concave Structures. IEICE Transactions on Information and Systems, J73-D-II(7): 992–1000.
3. S. Hirano and S. Tsumoto (2003): An Indiscernibility-Based Clustering Method with Iterative Refinement of Equivalence Relations - Rough Clustering -. Journal of Advanced Computational Intelligence and Intelligent Informatics, 7(2):169–177.
4. A. Yamada, Y. Shirai, and J. Miura (2002): Tracking Players and a Ball in Video Image Sequence and Estimating Camera Parameters for 3D Interpretation of Soccer Games. Proceedings of the 16th International Conference on Pattern Recognition (ICPR-2002), 1:303–306.
5. Y. Gong, L. T. Sin, C. H. Chuan, H. Zhang, and M. Sakauchi (1995): Automatic Parsing of TV Soccer Programs. Proceedings of the International Conference on Multimedia Computing and Systems (ICMCS'95), 167–174.
6. T. Taki and J. Hasegawa (2000): Visualization of Dominant Region in Team Games and Its Application to Teamwork Analysis. Computer Graphics International (CGI'00), 227–238.
7. T. Lindeberg (1990): Scale-Space for Discrete Signals. IEEE Trans. PAMI, 12(3), 234–254.
8. B. S. Everitt, S. Landau, and M. Leese (2001): Cluster Analysis Fourth Edition. Arnold Publishers.

Mobile Phone Data for Inferring Social Network Structure

Nathan Eagle[†], Alex (Sandy) Pentland[Δ] and David Lazer[*]

[†] nathan@mit.edu, MIT Design Laboratory, Massachusetts Institute of Technology, Cambridge, MA

[Δ] sandy@media.mit.edu, MIT Media Laboratory, Massachusetts Institute of Technology, Cambridge, MA

[*] david_lazer@harvard.edu, John F. Kennedy School of Government, Harvard University, Cambridge, MA

Abstract We analyze 330,000 hours of continuous behavioral data logged by the mobile phones of 94 subjects, and compare these observations with self-report relational data. The information from these two data sources is overlapping but distinct, and the accuracy of self-report data is considerably affected by such factors as the recency and salience of particular interactions. We present a new method for precise measurements of large-scale human behavior based on contextualized proximity and communication data alone, and identify characteristic behavioral signatures of relationships that allowed us to accurately predict 95% of the reciprocated friendships in the study. Using these behavioral signatures we can predict, in turn, individual-level outcomes such as job satisfaction.

1 Introduction

In a classic piece of ethnography from the 1940s, William Whyte carefully watched the interactions among Italian immigrants on a street corner in Boston's North End [1]. Technology today has made the world like the street corner in the 1940s—it is now possible to make detailed observations on the behavior and interactions of massive numbers of people. These observations come from the increasing number of digital traces left in the wake of our actions and interpersonal communications. These digital traces have the potential to revolutionize the study of collective human behavior. This study examines the potential of a particular device that has become ubiquitous over the last decade—the mobile phone—to collect data about human behavior and interactions, in particular from face-to-face interactions, over an extended period of time.

The field devoted to the study of the system of human interactions—social network analysis—has been constrained in accuracy, breadth, and depth because of its reliance on self-report data. Self-reports are potentially mediated by confounding factors such as beliefs about what constitutes a relationship, ability to recall interactions, and the willingness of individuals to supply accurate information about their relationships. Whole network studies relying on self-report relational data typically involve both limited numbers of people (usually less than 100) and a limited number of time points (usually 1). As a result, social network analysis has generally been limited to examining small, well-bounded populations, involving a small number of snapshots of interaction patterns [2]. While important work has been done over the last 30 years to parse the relationship between self-reported and observed behavior, much of social network research is written as if self-report data are behavioral data.

There is, however, a small but emerging thread of literature examining interaction data, e.g., based on e-mail [3,4] and call logs [5]. In this paper we use behavioral data collected from mobile phones [6] to quantify the characteristic behaviors underlying relational ties and cognitive constructs reported through surveys. We focus our study on three types of information that can be captured from mobile phones: communication (via call logs), location (via cell towers), and proximity to others (via repeated Bluetooth scans). The resulting data provide a multi-dimensional and temporally fine grained record of human interactions on an unprecedented scale. We have collected 330,000 hours of these behavioral observations from 94 subjects. Further, in principle, the methods we discuss here could be applied to hundreds of millions of mobile phone users.

2 Measuring Relationships

The core construct of social network analysis is the relationship. The reliability of existing measures for relationships has been the subject of sharp debate over the last 30 years, starting with a series of landmark studies in which it was found that behavioral observations were surprisingly weakly related to reported interactions [7, 8, 9]. These studies, in turn, were subject to three critiques: First, that people are far more accurate in reporting long term interactions than short term interactions [10]. Second, that it is possible to reduce the noise in network data because every dyad (potentially) represents two observations, allowing an evaluation (and elimination) of biases in the reports [11]. Third, that in many cases the construct of theoretical interest was the cognitive network, not a set of behavioral relations [12]. Here, behavior is defined as some set of activities that is at least theoretically observable by a third party, whereas a cognitive tie reflects some belief an individual holds about the relationship between two individuals [13].

There are multiple layers of conscious and subconscious cognitive filters that influence whether a subject reports a behavior [10, 14]. Cognitive sub-processes are engaged in the encoding and retrieval of a behavior instance from a subject's memory; the subject must understand the self-report request (i.e., survey question) to refer to the particular behavior; and the subject still gets to decide whether to report a particular behavior as a tie or not – a particular issue in the study of sexual or illicit relationships, for example [15]. These filtering processes contribute to a problematic gap between actual behaviors and self-report data.

Divergences between behavior and self-reports may be viewed as noise to be expunged from the data [11], or as reflecting intrinsically important information. For example, if one is interested in status, divergences between the two self-reports of a given relationship between two people, or between reported and observed behavior, may be of critical interest [16]. In contrast, if one is focused on the transmission of a disease, then the actual behaviors underlying those reports will be of central interest, and those divergences reflective of undesirable measurement error [15].

None of the above research examines the relationship between behavior and cognition for relationships that are intrinsically cognitive. Observing friendship or love is a fundamentally different challenge than observing whether two people talk to each other; e.g., two individuals can be friends without any observable interactions between them for a given period.

In this paper we demonstrate the power of collecting behavioral social network data from mobile phones. We first revisit the earlier studies on the inter-relationship between relational behavior and reports of relational behavior, but focusing in particular on some of the biases that the literature on memory suggest should arise. We then turn to the inter-relationship between behavior and reported friendships, finding that pairs of individuals that are friends demonstrate quite distinctive relational behavioral signatures. Finally, we show that these purely behavioral measures show powerful relationships with key outcomes of interest at the individual level—notably, satisfaction.

3 Research Design

This study follows ninety-four subjects using mobile phones pre-installed with several pieces of software that record and send the researcher data on call logs, Bluetooth devices in proximity, cell tower IDs, application usage, and phone status [17]. These subjects were observed via mobile phones over the course of nine months, representing over 330,000 person-hours of data (about 35 years worth of observations). Subjects included students and faculty from a major research institution; the resulting dataset is available for download. We also collected self-report relational data, where subjects were asked about their proximity to and friendship with others. Subjects were also asked about their satisfaction with their work group [18].

We conduct three analyses of these data. First, we examine the relationship between the behavioral and self-report interaction data. Second, we analyze whether there are behaviors characteristic of friendship. Third, we study the relationship between behavioral data and individual satisfaction.

4 Relationship between Behavioral and Self-Report Data

Subjects were asked how often they were proximate to other individuals at work. The boxplot shown in Figure 1 illustrates the remarkably noisy, if mildly positive, relationship between these self-report data and the observational data from Bluetooth scans. The literature on memory suggests a number of potential biases in the encoding into and retrieval from long term memory. We focus on two potential biases: recency and salience. Recency is simply the tendency for more recent events to be recalled [19]. Salience is the principle that prominent events are more likely to be recalled [20]. We therefore incorporate into our data analysis a measure of recent interactions (the week before the survey was answered), and a variety of measures of salience. The key question is whether recent and salient interactions significantly affect the subject's ability to accurately report average behaviors.

Using a multiple regression quadratic assignment procedure, common to the analysis of the adjacency matrices representing social networks, we can assess the significance of the predictive value of variables [18, 21]. While proximity at work was significantly related to self-reports, remarkably, proximity outside work was the single most powerful predictor of reported proximity at work. Other relational behavior, including proximity that was recent, on Saturday night, and between friends, were independently and significantly predictive of whether an individual reported proximity to someone else during work (p<.0001). These systematic biases limit the effectiveness of strategies designed to reduce noise in self-report data through modeling the biases of particular individuals [10], since these biases will affect both members of a dyad in the same direction (e.g., recency).

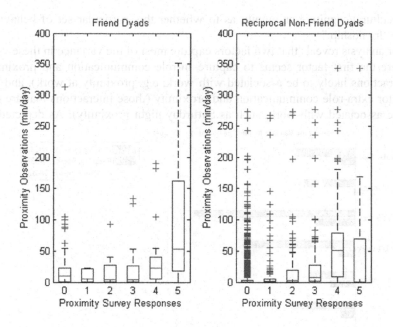

Figure 1. Self-Report vs. Observational Data. Boxplots highlighting the relationship between self-report and observational proximity behavior for undirected friendship and reciprocal non-friend dyads. Self-report proximity responses, on the x-axis, are scored from 0 to 5 (see legend). The y-axis shows observed proximity in minutes per day. The height of the box corresponds to the lower and upper quartile values of the distribution and the horizontal line corresponds to the distribution's median. The 'whiskers' extend from the box to values that are within 1.5 times the quartile range while outliers are plotted as distinct points. Three outlier dyads with an observed proximity greater than 400 min/day have been excluded from the plot.

What does a friendship "look like"? Certainly, we would anticipate relatively more phone calls and proximity between a pair of people who view one another as friends. More generally we anticipate that there are culturally embedded relational routines that friends tend to follow—for example, getting together outside of workplace hours and location, especially Saturday nights. We constructed seven dyadic behavioral variables: volume of phone communication and six contextualized variants of proximity. Figure 2 confirms that for all the dyadic behavioral variables, reciprocal friends score far higher than reciprocal non-friends (subjects who work together but neither considers the other a friend). A multivariate analysis confirms that the seven behavioral variables are significantly and independently related to reciprocated friendship/nonfriendship ($p < .001$). Further, in all but one case, non-reciprocal friends have intermediate scores. That one case is proximity at work, which suggests that

83

there is a cultural/cognitive ambiguity as to whether this particular set of behaviors constitutes "friendship."

A factor analysis reveals that two factors capture most of the variance in these variables, where the first factor seems to capture in-role communication and proximity (those interactions likely to be associated with work, e.g. proximity at work), and the second factor extra-role communication and proximity (those interactions that are unlikely to be associated with work, such as Saturday night proximity). As depicted in

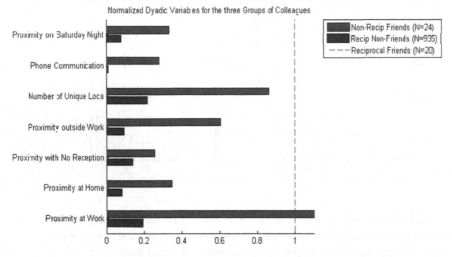

Figure 2. Normalized Dyadic Variables. The seven behavioral variables, normalized with respect to the reciprocal friendship data, are represented in the bar chart. The vertical dotted line at x=1 represents the values for reciprocal friend dyads. Reciprocal friends score higher than the other two groups for all dyadic variables with the exception of proximity at work. All three groups of dyads work together as colleagues.

Figure 3, a key finding of this study is that using just the extra-role communication factor from this analysis, it is possible to accurately predict 96% of symmetric non-friends and 95% of symmetric friends; in-role communication produces a similar accuracy. Thus we can accurately predict self reported friendships based only on objective measurements of behavior. These findings imply that the strong cultural norms associated with social constructs such as friendship produce differentiated and recognizable patterns of behavior. Leveraging these behavioral signatures to accurately characterize relationships in the absence of survey data has the potential to enable the quantification and prediction of social network structures on a much larger scale than is currently possible.

Unsurprisingly, non-reciprocal friendships fall systematically between these two categories. This probably reflects the fact that friendships are not categorical in na-

84

ture, and that non-reciprocal friendships may be indicative of moderately valued friendship ties. Thus, inferred friendships may actually contain more information than is captured by surveys that are categorical in nature. A pairwise analysis of variance using the Bonferroni adjustment shows that data from friendships, non-reciprocal friendships, and reciprocated non-friend relationships do indeed come from three distinct distributions ($F > 9$, $p < .005$).

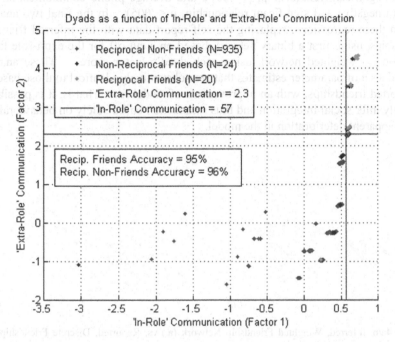

Figure 3. 'In-Role' Communication vs. 'Extra-Role' Communication. Each point represents a pair of colleagues' 'in-role' and 'extra-role' communication factor scores. 95% (19/20) of the reciprocal friendships have extra-role scores above 2.3, while 96% (901/935) of reciprocal non-friends have extra-role scores below 2.3.

6 Predicting Satisfaction Based on Behavioral Data

The preceding analysis highlights the potential to use the digital traces of previous behavior to infer cognitive constructs such as friendship. Do those inferences, in turn, predict meaningful individual level outcomes? One of the longest standing findings in the study of social support is the positive impact of social integration on the individual [22]. We examine here whether one can predict, in particular, satisfaction of the individual with their work group based solely on relational behavior. We begin with a standard analysis of the relationship between satisfaction and number of friends, which demonstrates a moderately positive and significant ($p < .05$), relationship. However, the model is significantly strengthened when we add two variables, combining self-report and behavioral data: average daily proximity to friends (a positive and significant relationship, $p < .001$), and average phone communication with friends (a negative and significant relationship, $p < .005$). In the final two analyses we reran these regressions, replacing the self-report data with the inferred friendship relationships, using first a binary network based on a cut off for the extra-role factor, and second, a weighted network using each dyad's factor score. These analyses produced a set of parameter estimates that are substantively identical to those based on self-reported friendships, with an improvement of model fit. That is, it is possible to accurately infer social integration and thus satisfaction based solely on behavioral data without apparent deterioration in the model.

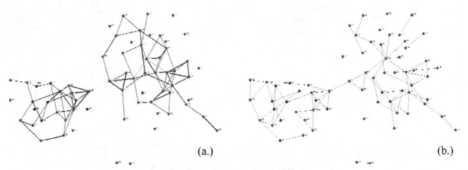

(a.)   (b.)

Figure 4a/b. Inferred, Weighted Friendship Network (a.) vs. Reported, Discrete Friendship Network (b.). The network on the left is the inferred friendship network with edge weights corresponding to the factor scores for factor 2, 'extra-role' communication. The network on the right is the reported friendship network. Node colors highlight the two groups of colleagues, first-year business school students (brown) and individuals working together in the same building (red).

7 Conclusions

This paper contains the results from a large scale study of physical proximity among individuals, encompassing 35 years worth of observations at five second increments, and combining them with phone log, locational, and self-report data. We anticipate that the methods outlined here will have a major impact in the social sciences, providing insight into the underlying relational dynamics of organizations, communities and, potentially, societies. At the micro level these methods, for example, provide a new approach to studying collaboration and communication within organizations—allowing the examination of the evolution of relationships over time. More dramatically, these methods allow for an inspection of the dynamics of macro networks that were heretofore unobservable. There is no technical reason why data cannot be collected from hundreds of millions of people throughout the course of their lives. Further, while the collection of such data raises serious privacy issues that need to be considered, the potential for achieving important societal goals is considerable. The implications for epidemiology alone are foundational, as they are for the study of sociology, politics, and organizations, among other social sciences.

This paper thus offers a necessary first step in this revolution, linking the predominant existing methodologies to collect social network data, based on self reports, to data that can be collected automatically via mobile phones. Our results suggest that behavioral observations from mobile phones provide insight not just into observable behavior, but also into purely cognitive constructs, such as friendship and individual satisfaction. While the specific results are surely embedded within the social milieu in which the study was grounded, the critical next question is how much these patterns vary from context to context.

References

1. W. Whyte. 1993. Street Corner Society: The Social Structure of an Italian Slum. 4th edition. Chicago, IL: University of Chicago Press.
2. S. Wasserman and K. Faust. 1994. Social Network Analysis, Methods and Applications. Cambridge, UK: Cambridge University Press.
3. G. Kossinets and D. J. Watts. 2006. "Empirical Analysis of an Evolving Social Network," Science 311: 88-90.
4. H. Ebel, L-I. Mielsch, and S. Bornholdt. 2002. "Scale-free topology of e-mail networks," Phys Rev E 66: 35103.
5. W. Aiello, F. Chung, L. Lu. 2000. A random graph model for massive graphs, Annual ACM Symposium on Theory of Computing, Proceedings of the thirty-second annual ACM symposium on Theory of computing: 171–180.
6. N. Eagle, A. Pentland. 2006. "Reality Mining: Sensing Complex Social Systems", Personal and Ubiquitous Computing, 10(4): 255-268.
7. W. H. Bernard, P. Killworth, and L. Sailer. 1979. "Informant accuracy in social networks. Part IV: A comparison of clique-level structure in behavioral and cognitive network data." Social Networks 2: 191-218.
8. P. D. Killworth and H. R. Bernard. 1976. "Informant accuracy in social network data," Human Organization 35(8): 269-286.
9. P.V. Marsden. 1990. "Network data and measurement," Annual Review of Sociology 16: 435-463.

10. L.C. Freeman, A.K. Romney, S.C. Freeman. 1989. "Cognitive Structure and Informant Accuracy," American Anthropologist 89.
11. C. Butts. 2003. "Network inference, error and informant (in)accuracy: a Bayesian approach," Social Networks 25:2: 103-140.
12. W.D. Richards. 1985. "Data, models, and assumptions in network analysis. In Organizational Communication: Traditional Themes and New Directions, ed R. D. McPhee , P. K. Tompkins. Sage, Beverly Hills: 108-128.
13. D. Krackhardt. 1990. "Assessing the Political Landscape: Structure, Power, and Cognition in Organizations." Administrative Science Quarterly, 35: 342–369.
14. L.C. Freeman. 1992. Filling in the blanks: a theory of cognitive categories and the structure of social affiliation. Social Psychology Quarterly 55(2): 118-127.
15. Brewer, D. D., Potterat, J. J., Muth, S. Q., Malone, P. Z., Montoya, P. A., Green, D. A., Rogers, H. L., & Cox, P. A. 2005. "Randomized trial of supplementary interviewing techniques to enhance recall of sexual partners in contact interviews. Sexually Transmitted Diseases, 32, 189-193.
16. K.M. Carley and D. Krackhardt. 1996. "Cognitive inconsistencies and non-symmetric friendships." Social Networks 15: 377-398.
17. M. Raento, A. Oulasvirta, R. Petit,H. Toivonen. 2005. "ContextPhone – A prototyping plat-form for context-aware mobile applications". IEEE Pervasive Computing, 4 (2), 51-59.
18. Full details on data collection and variable construction are available in the supporting online materials.
19. Waugh, N. C., & Norman, D. A. 1965. Primary memory. Psychological Review 72: 89-104.
20. Higgins, E. T. 1996. Knowledge activation: Accessibility, applicability, and salience. In E. T. Higgins & A. Kruglanski (Eds.), Social psychology: Handbook of basic principles (pp.133–168). New York: Guilford Press.
21. Krackhardt, D. 1988. Predicting with Networks - Nonparametric Multiple-Regression Analysis of Dyadic Data. Social Networks 10 (4): 359-81.
22. Durkheim, Emile. 1951. Suicide: A Study in Sociology translated by George Simpson and John A. Spaulding. New York: The Free Press.

Human Behavioral Modeling Using Fuzzy and Dempster–Shafer Theory

Ronald R. Yager

yager@panix.com, Machine Intelligence Institute, Iona College, New Rochelle, NY

Abstract: Human behavioral modeling requires an ability to represent and manipulate imprecise cognitive concepts. It also needs to include the uncertainty and unpredictability of human action. We discuss the appropriateness of fuzzy sets for representing human centered cognitive concepts. We describe the technology of fuzzy systems modeling and indicate its the role in human behavioral modeling. We next introduce some ideas from the Dempster-Shafer theory of evidence. We use the Dempster-Shafer theory to provide a machinery for including randomness in the fuzzy systems modeling process. This combined methodology provides a framework with which we can construct models that can include both the complex cognitive concepts and unpredictability needed to model human behavior.

1 Human Behavioral Modeling

Two important classes of human behavioral modeling can be readily identified. The first is the modeling of some physical phenomenon or system involving human participants. This is very much what is done in social sciences and is clearly inspired by the classical successful use of modeling in physics and engineering. The modeling here is from the perspective of an external observer. We can refer to this as E-O modeling. The second type of modeling, of much more recent vintage, can be denoted as I-P modeling as an acronym for Internal Participant modeling. This type of modeling has arisen to importance with the wide spread use digital technology. It is central in the construction of synthetic agents, computational based training systems and machine learning. It is implicit in our attempts to construct intelligent systems. Here we are trying to digitally model a "human" or "human like" agent that interacts with some more complex environment which itself can be digital or real or some combination.

In either case, and perhaps more so in the I-P situation, human behavioral modeling requires an ability to formally represent sophisticated cognitive concepts that are often at best described in imprecise linguistic terms. Set based methods and more particularly fuzzy sets provide a powerful tool for enabling the semantical modeling of these imprecise concepts within computer based systems [1-2]. With the aid of a fuzzy set we can formally represent sophisticated imprecise linguistic concepts in a manner that allows for the types of computational manipulation needed for reasoning in behavioral models based on human cognition and conceptualization. Central to the use of fuzzy sets is the ability to capture the "grayness" of human conceptualization. Most concepts used in human behavioral modeling, both from the E-O and I-P perspective, are not binary but gradually go from clearly yes to clearly no. Furthermore in discussing the qualities of important social relationships such as political ties, kinship obligations and friendship we use attributes such as intensity, durability and reciprocity [3]. These attributes most naturally evaluated in imprecise terms. In modeling the rules determining the behavior of some simulated agent we must have the ability to model the kinds fluidity central to the human capacity to adapt and deal with new situations.

Fuzzy systems modeling (FSM) [4] is a rule based technique that allows for formal reasoning and manipulation with the types of imprecise concepts central to human cognition. It can use a semantic understanding of an age related concept such as *old* in order to be able how well a particular individual satisfies the concept.. Clearly FSMs can be used to model the types of complex relationships needed in human behavioral modeling. It is the basic technique used in the development of many successful applications [5]. FSM helps simply the task of modeling complex relationship and processes by partitioning the input (antecedent) space into regions in which one can more easily comprehend and express the appropriate consequents. In FSM the rules are expressed in linguistic terms with a representation using fuzzy subsets. An important feature of the FSM is that it can create and formulate new solutions. That is the output of an FSM does not have to be one of the consequents of a rule but can be constructed out of a combination of outputs from different rules.

In addition to the imprecision of human conceptualization reflected in language many situations that arise in human behavioral modeling entail aspects of probabilistic uncertainty. This is true in both E-O and I-P applications. Consider an observation such as "Generally women of child bearing age do not get to close to foreigners" Here we see imprecise terms such as "child bearing age" and "close" as well as the term "generally" conveying a probabilistic aspect. In this this work we describe a methodology for including probabilistic uncertainty in the fuzzy systems model. The technique we suggest for the inclusion of this uncertainty is based upon the Dempster-Shafer theory of evidence [6, 7]. The Dempster-Shafer approach fits nicely into the FSM technique since both techniques use sets as their primary data structure and are important components of the emerging field of granular computing [8, 9].

We first discuss the fundamentals of FSM based on the Mamdani reasoning paradigm. We next introduce some of the basic ideas from the Dempster-Shafer theory which are required for our procedure. We then show how probabilistic uncertainty in the output of a rule based model can be included in the FSM using the Dempster-Shafer (D-S) paradigm. We described how various types of uncertainty can be modeled using this combined FSM / D-S paradigm.

2 Fuzzy Systems Modeling

Fuzzy systems modeling (FSM) provides a technology for the development of semantically rich rule based representations that can model complex, nonlinear multiple input output relationships or functions or systems.

The technique of FSM allows one to represent the model of a system by partitioning the input space. Thus if $U_1, \ldots U_r$ are the input (antecedent) variables and V is the output (consequent) variable we can represent their relationship by a collection n of "rules" of the form,

When U_1 is A_{i1} and U_2 is $A_{i2}, \ldots$ and U_r is A_{ir} then V is D_i.

Here each A_{ij} typically indicates a linguistic term corresponding to a value of its associated variable, for example if U_j is the variable correspond to age then A_{ij} could be "young." or "child bearing age." Furthermore each A_{ij} is formally represented as a fuzzy subset over the domain X_j of the associated variable U_j Similarly D_i is a value associated with the consequent variable V that is formally defined as a fuzzy subset of the domain Y of V.

In the preceding rules the antecedent specifies a condition that if met allows us to infer that the possible value for the variable V lies in the consequent subset D_i. For each rule the antecedent defines a fuzzy region of the input space, $X_1 \times X_2 \times ... \times X_m$, such that if the input lies in this region the consequent holds. Taken as a collection the antecedents of all the rules form a fuzzy partition of the input space. A key advantage of this approach is that by partitioning the input space we can allow simple functions to represent the consequent.

The process of finding the output of a fuzzy systems model for given values of the input variables is called the "reasoning" process. One method for reasoning with fuzzy systems models is the Mamdani-Zadeh paradigm. [10].

Assume the input to a FSM consists of the values $U_j = x_j$. In the following we shall use the notation $A_{ij}(x_j)$ to indicate the membership of the element x_j in the fuzzy subset A_{ij}. This can be seen as the degree of truth of the proposition U_j is A_{ij} given that $U_j = x_j$. The procedure for reasoning used in the Mamdani-Zadeh method consists of the following steps:

1. For each rule calculate its firing level $\tau_i = \text{Min}_j[A_{ij}(x_j)]$

2. Calculate the output of each rule as a fuzzy subset F_i of Y where

$$F_i(y) = \text{Min}[\tau_i, D_i(y)]$$

3. Aggregate the individual rule outputs to get a fuzzy subset F and Y where
$$F(y) = \text{Max}_i[F_i(y)].$$

F is a fuzzy subset of Y indicating the output of the system. It is important to emphasize that F can be something new, it has been constructed from distinct components of the rule base.

At this point we can describe three options with respect to presenting this output to the final user. The simplest is to give them the fuzzy set F. This of course is the least appealing especially if the user is not technically oriented. The second, and perhaps the most sophisticated, is to perform what is called retranslation. Here we try to express the fuzzy set F in some kind appropriate linguistic form. While we shall not pursue this approach here we note that in [11] we have investigated the process of retranslation. The third alternative is to compress the fuzzy set F into some precise value from the space Y. This process is called defuzzification. A number techniques are available to implement the defuzzification. Often the choice is dependent upon the structure of the space Y associated with variable V. One approach is to take as the output the element in Y that has the largest membership in F. While available in most domains it loses a lot of the information. A preferred approach, if the under lying structure of Y allows, is to take a kind of weighted average using the membership grades in F to provide the weights. The most commonly used procedure for defuzzification process is the center of gravity. Using this method we calculate the

defuzzification value as $\bar{y} = \dfrac{\sum_i y_i F(y_i)}{\sum_i F(y_i)}$.

3 Dempster-Shafer Theory of Evidence

In this section we introduce some ideas from the Dempster-Shafer uncertainty theory [6, 7]. Assume X is a set of elements. Formally a Dempster-Shafer belief structure m is a collection of q non-null subsets A_i of X called focal elements and a set of associated weights $m(A_i)$ such that: (1) $m(A_i) \in [0, 1]$ and (2) $\sum_i m(A_i) = 1$.

One interpretation that can be associated with this structure is the following. Assume we perform a random experiment which can have one of q outcomes. We shall denote the space of the experiment as Z. Let P_i be the probability of the ith outcome z_i. Let V be another variable taking its value in the set X. It is the value of the variable V that is of interest to us. The value of the variable V is associated with the performance of the experiment in the space Z in the following manner. If the outcome of the experiment on the space Z is the ith element, z_i, we shall say that the value of V lies in the subset A_i of X. Using this semantics we shall denote the value of the variable as V is m, where m is a Dempster-Shafer granule with focal elements A_i and weights $m(A_i) = P_i$.

A situation which illustrates the above is the following. We have three candidates for president, Red, White and Blue. The latest polling information indicates that the probabilities of each candidate winning is (Red, 0.35), (White, 0.55) and (Blue, 0.1). Our interest here is not on who will be president but on the future interest rates. Based on the campaign statements of the three candidates we are able to conclude that Red will support low interest rates and White will support $high$ interest rates. For the candidate

Blue we no have information about his attitude toward interest rates. The Dempster-Shafer framework provides an ideal structure for representing this knowledge. Here we let V be the variable corresponding to the future interest rates and let X be the set corresponding to the domain of interest rates, the variable V will assume its value in X. We can now represent our knowledge the value of the future interest rates V using the Dempster-Shafer framework. Here we have three focal sets. The first, A_1, is "low interest rates." The second, A_2 is "high interest rates." The third, A_3, is "unknown interest rate." Furthermore the associated weights are $m(A_1) = 0.35$, $m(A_2)$ = 0.55 and $m(A_3) = 0.1$. Each of the A_j are formulated as subsets of X. We note A_3, "unknown interest rate, " is the set X.

Here our interest is in finding the probabilities of events associated with V, that is with arbitrary subsets of X. For example we may be interested in the probability that interest rates will be *less then 4 %*. Because of the imprecision in the information we can't find exact probabilities but we must settle for ranges. Two measures are introduced to capture the relevant information.

Let B be a subset of X the **plausibility** of B, denoted Pl(B), is defined as

$$Pl(B) = \sum_{i, A_i \cap B \neq \emptyset} m(A_i),$$

The **belief** of B, denoted Bel(B), is defined as

$$Bel(B) = \sum_{i, B \subseteq A_i} m(A_i).$$

For any subset B of X $Bel(B) \leq Prob(B) \leq Pl(B)$. The plausibility and belief provide upper and lower bounds on the probability of the subset B.

An important issue in the theory of Dempster-Shafer is the procedure for aggregating multiple belief structures on the same variable. This can be seen as a problem of information fusion. This standard procedure is called Dempster's rule, it is a kind of conjunction (intersection) of the belief structures.

Assume m_1 and m_2 are two independent belief structures on the space X their conjunction is another belief structure m, denoted $m = m_1 \oplus m_2$. The belief structure m is obtained in the following manner. Let m_1 have focal elements A_i, $i = 1$ to n_1 and let m_2 have focal elements B_j, $j = 1$ to n_2. The focal elements of m are all the subsets $F_K = A_i \cap B_j \neq \emptyset$ for some i and j. The associated weights are $m(F_K) = \dfrac{1}{1-T} (m_1(A_i) *$

$m_2(B_j)$ where $T = \sum_{A_i \cap B_j = \emptyset} m_1(A_i) * m_2(B_j)$.

Example: Assume our universe of discourse is X = {1, 2, 3, 4, 5, 6}

m1		m2	
$A_1 = \{1, 2, 3\}$	$m_1(A_1) = 0.5$	$B_1 = \{2, 5, 6\}$	
	$m_2(B_1) = 0.6$		
$A_2 = \{2, 3, 6\}$	$m_1(A_2) = 0.3$	$B_2 = \{1, 4\}$	
	$m_2(B_2) = 0.4$		

$A_3 = \{1, 2, 3, 4, 5, 6\}$ $m_1(A_3) = 0.2$

Taking the conjunction we get: $F_1 = A_1 \cap B_1 = \{2\}$, $F_2 = A_1 \cap B_2 = \{1\}$, $F_3 = A_2 \cap B_1 = \{2, 6\}$, $F_4 = A_3 \cap B_1 = \{2, 5, 6\}$ and $F_5 = A_3 \cap B_1 = \{1, 4\}$.

We note that $A_2 \cap B_2 = \varnothing$. Since only one intersection gives us the null set then $T = m_1(A_2) * m(B_2) = .12$ and $1 - T = 0.88$. Using this we get $m(F_1) = 0.341$, $m(F_2) = 0.227$, $m(F_3) = 0.205$, $m(F_4) = 0.136$ and $m(F_5) = 0.09$.

The above combination of belief structures can be seen to be essentially an intersection, conjunction, of the two belief structures. In [12] Yager provided for an extension of the aggregation of belief structures to any set based operation. Assume ∇ is any binary operation defined on sets, $D = A \nabla B$ where A, B and D are sets. We shall say that ∇ is an "non-null producing" operator if $A \nabla B \neq \varnothing$ when $A \neq \varnothing$ and $B \neq \varnothing$. The union is non-null producing but intersection is not. Assume m_1 and m_2 are two belief structures with focal elements A_i and B_j respectively. Let ∇ be any non-null producing operator. We now define the new belief structure $m = m_1 \nabla m_2$. The belief structure m has focal elements $E_K = A_i \nabla B_j$ with $m(E_K) = m_1(A_i) * m_2(B_j)$. If ∇ is **not** non-null producing we may be forced to do a process called normalization [12]. The process of normalization consists of the following

(1) Calculate $T = \displaystyle\sum_{A_i \nabla B_j = \varnothing} m_1(A_i) * m(B_j)$

(2) For all $E_K = A_i \nabla B_j \neq \varnothing$ calculate $m(E_K) = \dfrac{1}{1 - T} m_1(A_i) * m_2(B_j)$

(3) For all $E_K = \varnothing$ set $m(E_K) = 0$.

We can use the Dempster-Shafer structure to represent some very naturally occurring types of information. Assume V is a variable taking its value in the set X. Let A be a subset of X. Assume our knowledge about V is that the probability that V lies in A is "at least α." This information can be represented as the belief structure m which has two focal elements A and X and where $m(A) = \alpha$ and $m(X) = 1$. The information that the probability of A is exactly α can be represented as a belief structure m with focal elements A and $\overline{A}$ where $m(A) = \alpha$ and $m(\overline{A}) = 1 - \alpha$.

An ordinary probability distribution P can also be represented as a belief structure. Assume for each element $x_i \in X$ it is the case P_i is its probability. We can represent this as a belief structure where the focal elements are the individual element $A_i = \{x_i\}$ and $m(A_i) = P_i$. For these types of structures it is the case that for any subset A of X, $Pl(A) = Bel(A)$, thus the probability is uniquely defined as a point rather than interval.

The D-S belief structure can be extended to allow for fuzzy sets [13, 14]. To extend the measures of plausibility and belief we need two ideas from the theory of possibility [15]. Assume A and B are two fuzzy subsets of X, the possibility of B given A is defined as $Poss[B/A] = Max_i[A(x_i) \wedge B(x_i)]$ where $\wedge$ is the min. The certainty of B given A is $Cert[B/A] = 1 - Poss[\overline{B}/A]$. Here $\overline{B}$ is the complement of B, it has membership grade $\overline{B}(x) = 1 - B(x)$.

Using these we extend the concepts of plausibility and belief. If m is a belief structure on X with focal fuzzy elements A_i and B is any fuzzy subset of X then $Pl(B) = \sum_i Poss[B/A_i] \, m(A_i)$ and $Bel(B) = \sum_i Cert[B/A_i] \, m(A_i)$. The plausibility and belief measures are the expected possibility and certainty of the focal elements.

The combination of belief structures with fuzzy focal elements can be made. If ∇ is some set operation we simply use the fuzzy version of it. For example if m_1 and m_2 are belief structures with fuzzy focal elements then $m = m_1 \cup m_2$ has focal elements $E_K = A_i \cup B_j$ where $E_K(x) = A_i(x) \vee B_j(x)$ ($\vee$ = max). Here as in the non-fuzzy case $m(E_K) = m_1(A_i) \, m_2(B_j)$.

Implicit in the formulation for calculating the new weights is an assumption of independence between the belief structures. This independence is reflected in an assumption that the underlying experiments generating the focal elements for each belief structure are independent. This independence manifests itself in the use of the product to calculate the new weights. That is the joint occurrence of the pair of focal elements A_i and B_j is the product of probabilities of each of them $m_1(A_i)$ and $m_2(B_j)$.

In some situations we may have a different relationship between the two belief structures. One very interesting case is called **synonymity**. For two belief structures to be in synonymity they must have their focal elements induced from the same experiment. Thus if m_1 and m_2 are two belief structures on X that are in synonymity they should have the same number of focal elements with the same weights. Thus the focal elements of m_1 are A_i for i = 1 to q, and those of m_2 are are B_j for i = 1 to q then $m_1(A_i) = m_2(B_j)$. In the case of synonymity between m_1 and m_2 if ∇ is any non-null producing set operator then $m = m_1 \nabla m_2$ also has n focal elements $E_i = A_i \nabla B_i$ with $m(E_i) = m(A_i) = m(B_i)$.

4 Probabilistic Uncertainty in the FSM

In the basic FSM, the Mamdani-Zadeh model, the consequent of each rule consists of a fuzzy subset. The consequent of an individual rule is a proposition of the form V *is* D_i. The use of a fuzzy subset implies a kind of uncertainty associated with the output of a rule. The kind of uncertainty is called possibilistic uncertainty and is a reflection of a lack of precision in describing the output. The intent of this proposition if to indicate that the value of the output is constrained by (lies in) the subset D_i.

We now shall add further modeling capacity to the FSM technique by allowing for probabilistic uncertainty in the consequent. A natural extension of the FSM is to consider the consequent to be a fuzzy Dempster-Shafer granule. Thus we shall now consider the output of each rule to be of the form V *is* m_i where m_i is a belief structure with focal elements D_{ij} which are fuzzy subsets of the universe Y and associated weights $m_i(D_{ij})$. Thus a typical rule is now of the form

When U_1 is A_{i1} and U_2 is $A_{i2}, \ldots U_r$ is A_{ir} then V is m_i.

Using a belief structure to model the consequent of a rule is essentially saying that $m_i(D_{ij})$ is the probability that the output of the ith rule lies in the set D_{ij}. So rather than being certain as to the output set of a rule we have some randomness in the rule. We note that with $m_i(D_{ij}) = 1$ for some D_{ij} we get the original FSM.

We emphasize that the use of a fuzzy Dempster-Shafer granule to model the consequent of a rule brings with it two kinds of uncertainty. The first type of uncertainty is the randomness associated with determining which of the focal elements of m_i is in effect if the rule fires. This selection is essentially determined by a random experiment which uses the weights as the appropriate probability. The second type of uncertainty is related to the selection of the outcome element given the fuzzy subset, this is related to the issue of lack of specificity. This uncertainty is essentially resolved by the defuzzification procedure used to pick the crisp singleton output of the system.

We now describe the reasoning process in this situation with belief structure consequents. Assume the input to the system are the values for the antecedent variables, $U_j = x_j$. The process for obtaining the firing levels of the individual based upon these inputs is exactly the same as in the previous situation.

For each rule we obtain the firing level, $\tau_i = \text{Min}[A_{ij}(x_j)]$.

The output of each rule is a belief structure $\widehat{m}_i = \tau_i \wedge m$. The focal elements of $\widehat{m}_i$ are F_{ij} a fuzzy subset of Y where $F_{ij}(y) = \text{Min}[\tau_i, D_{ij}(y)]$, here D_{ij} is a focal element of m_i. The weights associated with these new focal elements are simply $\widehat{m}_i(F_{ij}) = m_i(D_{ij})$.

The overall output of the system m is obtained in a manner analogous to that used in the basic FSM, we obtain m by taking a union of the individual rule outputs, $m = \bigcup\limits_{i=1}^{n} \widehat{m}_i$.

Earlier we discussed the process of taking the union of belief structures. For every a collection $<F_{1j_1}, \ldots F_{nj_n}>$ where F_{ij_i} is a focal element of $\widehat{m}_i$ we obtain a focal element of m, $E = \bigcup\limits_{i} F_{ij_i}$ and the associated weight is $m(E) = \prod\limits_{i=1}^{n} \widehat{m}_i(F_{ij_i})$.

As a result of this third step we obtain a fuzzy D-S belief structure V is m as our output of the FSM. We denote the focal elements of m as the fuzzy subsets $E_j, j = 1$ to q, with weights $m(E_j)$. Again we have three choices: present this to a user, try to linguistically summarize the belief structure or to defuzzify to a single value. We shall here discuss the third option.

The procedure used to obtain this defuzzified value $\overline{y}$ is an extension of the previously described defuzzification procedure. For each focal element E_j we calculate its defuzzified value $\bar{y}_j = \dfrac{\Sigma_i y_i \, E_j(y_i)}{\Sigma_i E_j(y_i)}$. We then obtain as the defuzzified value of m, $\bar{y} =$

$\Sigma_j \bar{y}_j \, m(E_j)$. Thus $\bar{y}$ is the expected defuzzified value of the focal elements of m.

The following simple example illustrates the technique just described.

Example: Assume a FSM has two rules

$$\text{If U } is \text{ A}_1 \text{ then V } is \text{ m}_1$$
$$\text{If U } is \text{ A}_2 \text{ then V } is \text{ m}_2.$$

m_1: has focal elements D_{11} = "about two" = $\left\{\dfrac{.6}{1}, \dfrac{1}{2}, \dfrac{.6}{3}\right\}$ and D_{12} = "about five" =

$\left\{\dfrac{.5}{4}, \dfrac{1}{5}, \dfrac{.6}{6}\right\}$ with $m_1(D_{11}) = 0.7$ and $m_1(D_{12}) = 0.3$.

m_2: has focal elements D_{21} = "about 10" = $\left\{\dfrac{.7}{9}, \dfrac{1}{10}, \dfrac{.7}{11}\right\}$ and D_{22} = "about 15" =

$\left\{\dfrac{.4}{14}, \dfrac{1}{15}, \dfrac{.4}{10}\right\}$ with $m_2(D_{21}) = 0.6$ and $m_2(D_{22}) = 0.4$

Assume the system input is x* and the membership grade of x* in A_1 and A_2 are 0.8 and 0.5 respectively. Thus the firing levels of each rule are $\tau_1 = 0.8$ and $\tau_2 = 0.5$. We now calculate the output each rule $\widehat{m}_1 = \tau_1 \wedge m_1$ and $\widehat{m}_2 = \tau_2 \wedge m_2$.

$\widehat{m}_1$: has focal elements $F_{11} = \tau_1 \wedge D_{11} = \left\{\dfrac{.6}{1}, \dfrac{.8}{2}, \dfrac{.6}{3}\right\}$ and $F_{12} = \tau_1 \wedge D_{12} =$

$\left\{\dfrac{.5}{4}, \dfrac{.8}{5}, \dfrac{.6}{6}\right\}$ with $m(F_{11}) = 0.7$ and $m(F_{12}) = 0.3$

$\widehat{m}_2$: has focal elements $F_{21} = \tau_2 \wedge D_{21} = \left\{\dfrac{.5}{9}, \dfrac{.5}{10}, \dfrac{.5}{11}\right\}$ and $F_{22} = \tau_2 \wedge D_{22} =$

$\left\{\dfrac{.4}{14}, \dfrac{.5}{15}, \dfrac{.4}{10}\right\}$ with $m(F_{21}) = 0.6$ and $m(F_{22}) = 0.4$

We next obtain the union of these two belief structure, m = $m_1 \cup m_2$ with focal elements

$$E_1 = F_{11} \cup F_{21} \qquad m(E_1) = \widehat{m}_1(F_{11}) * \widehat{m}_2(F_{21})$$
$$E_2 = F_{11} \cup F_{22} \qquad m(E_2) = \widehat{m}_1(F_{11}) * \widehat{m}_2(F_{22})$$
$$E_3 = E_{12} \cup F_{21} \qquad m(E_3) = \widehat{m}_1(F_{12}) * \widehat{m}_2(F_{21})$$
$$E_4 = E_{12} \cup F_{22} \qquad m(E_4) = \widehat{m}_1(F_{12}) * \widehat{m}_2(F_{22})$$

Doing the above calculations we get

$$E_1 = \left\{\dfrac{0.6}{1}, \dfrac{0.8}{2}, \dfrac{0.6}{3}, \dfrac{0.5}{9}, \dfrac{0.5}{10}, \dfrac{0.5}{11}\right\} \qquad\qquad m(E_1) \quad =$$

0.42

$$E_2 = \left\{\dfrac{0.6}{1}, \dfrac{0.8}{2}, \dfrac{0.6}{3}, \dfrac{0.4}{14}, \dfrac{0.5}{15}, \dfrac{0.4}{10}\right\} \qquad\qquad m(E_2) \quad =$$

0.28

$$E_3 = \left\{\dfrac{0.5}{4}, \dfrac{0.8}{5}, \dfrac{0.6}{6}, \dfrac{0.5}{9}, \dfrac{0.5}{10}, \dfrac{0.5}{11}\right\} \qquad\qquad m(E_3) \quad =$$

0.18

$$E_4 = \left\{ \frac{0.5}{4}, \frac{0.8}{5}, \frac{0.6}{6}, \frac{0.4}{14}, \frac{0.5}{15}, \frac{0.4}{10} \right\} \qquad m(E_4) = 0.12$$

We now proceed with the defuzzification of the focal elements.
Defuzzy(E_1) = $\bar{y}_1$ =5.4, Defuzzy(E_2) = $\bar{y}_2$ = 6.4, Defuzzy(E_3) = $\bar{y}_3$ = 7.23 and Defuzzy(E_4) = $\bar{y}_4$ = 8.34. Finally taking the expected value of these we get

$$\bar{y} = (0.42)(5.4) + (0.28)(6.4) + (0.18)*(7.23) + (0.12)(8.34) = 6.326$$

The development of FSMs with Dempster-Shafer consequents allows for the representation of different kinds of uncertainty associated with the modeling rules.

One situation is where we have a value $\alpha_i \in [0, 1]$ indicating the confidence we have in the ith rule. In this case we have a nominal rule of the form

If U *is* A_i then V *is* B_i

with confidence "at least α_i".

Using the framework developed above we can transform this rule, along with its associated confidence level into a Dempster-Shafer structure

"If U *is* A_i then V *is* m_i."

Here m_i is a belief structure with two focal elements, B_i and Y. We recall Y is the whole output space. The associated weights are $m_i(A_i) = \alpha_i$ and $m(Y) = 1 - \alpha_i$. We see that if $\alpha_i = 1$ then we get the original rule while if $\alpha_i = 0$ we get a rule of the form

If U is A_i then V is Y.

5 Conclusion

We have suggested a framework which can be used for modeling human behavior. The approach suggested has the ability to represent the types of linguistically expressed concepts central to human cognition. It also has a random component which enables the modeling of the unpredictability of human behavior.

References

1. Zadeh, L. A., "A note on web intelligence, world knowledge and fuzzy logic," Data and Knowledge Engineering 50, 291-304, 2004.
2. Yager, R. R., "Using knowledge trees for semantic web querying," in Fuzzy Logic and the Semantic Web, edited by Sanchez, E., Elsevier: Amsterdam, 231-246, 2006.
3. Scott, J., Social Network Analysis, SAGE Publishers: Los Angeles, 2000.
4. Pedrycz, W. and Gomide, F., Fuzzy Systems Engineering: Toward Human-Centric Computing, John Wiley & Sons: New York, 2007.
5. Sugeno, M., Industrial Applications of Fuzzy Control, North-Holland: Amsterdam, 1985.

6. Shafer, G., A Mathematical Theory of Evidence, Princeton University Press: Princeton, N.J., 1976.
7. Yager, R. R. and Liu, L., (A. P. Dempster and G.Shafer, Advisory Editors) Classic Works of the Dempster-Shafer Theory of Belief Functions, Springer: Heidelberg, (To Appear).
8. Lin, T. S., Yao, Y. Y. and Zadeh, L. A., Data Mining, Rough Sets and Granular Computing, Physica-Verlag: Heidelberg, 2002.
9. Bargiela, A. and Pedrycz, W., Granular Computing: An Introduction, Kluwer Academic Publishers: Amsterdam, 2003.
10. Yager, R. R. and Filev, D. P., Essentials of Fuzzy Modeling and Control, John Wiley: New York, 1994.
11. Yager, R. R., "On the retranslation process in Zadeh's paradigm of computing with words," IEEE Transactions on Systems, Man and Cybernetics: Part B 34, 1184-1195, 2004.
12. Yager, R. R., "Arithmetic and other operations on Dempster-Shafer structures," Int. J. of Man-Machine Studies 25, 357-366, 1986.
13. Yager, R. R., "Entropy and specificity in a mathematical theory of evidence," Int. J. of General Systems 9, 249-260, 1983.
14. Yen, J., "Generalizing the Dempster-Shafer theory to fuzzy sets," IEEE Transactions on Systems, Man and Cybernetics 20, 559-570, 1990.
15. Dubois, D. and Prade, H., "Possibility theory as a basis for qualitative decision theory," Proceedings of the International Joint Conference on Artificial Intelligence (IJCAI), Montreal, 1924-1930, 1995.

Online Behavioral Analysis and Modeling Methodology (OBAMM)[1]

David J. Robinson[†], Vincent H. Berk[†], and George V. Cybenko[†]

[†]Firstname.M.Lastname@Dartmouth.EDU, Thayer School of Engineering at Dartmouth College, Hanover, New Hampshire

Abstract This paper introduces a novel method of tracking user computer behavior to create highly granular profiles of usage patterns. These profiles, then, are used to detect deviations in a users' online behavior, detecting intrusions, malicious insiders, misallocation of resources, and out-of-band business processes. Successful detection of these behaviors significantly reduces the risk of leaking sensitive data, or inadvertently exposing critical assets.

1 Introduction

The World Wide Web (WWW) has become the single largest repository of information in the world, with people utilizing it to address all aspects of their home and work lives. As our dependence on the web increases, so does the amount of information that can be gathered about an individual using the Internet. E-commerce and marketing firms have taken advantage of this fact for years by accumulating information on individuals for purposes ranging from tailoring ad campaigns to personalizing a shopping experience. Although of obvious importance to marketing firms, this data offers potential benefits in other areas as well. So far, in computer security and intellectual property protection, little attention has been paid to the area of user behavioral profiling. While profiling is generally frowned upon when dealing with personal privacy issues, the ability to accurately profile and depict user activity could eliminate many forms of computer related criminal activity. Using profiling information in the area of computer security can aid significantly in the detection of rogue users, malicious activity, policy violations and unauthorized data exfiltration, and in many cases will even prevent them from happening.

---

[1]   This work was supported under DHS contract number 2006-CS-001-000001. The views expressed in this work are those of the authors alone, and do not necessarily represent the official position of the US DHS.

100

In this paper, we suggest an Online Behavioral Analysis and Modeling Methodology (OBAMM) to accurately and efficiently categorize users based on their individual web browsing activities and propose how this information may be used in the realm of computer security.

Section 2 of the paper provides background on other approaches that take advantage of profiling information. Section 3 briefly describes OBAMM followed by initial results in Section 4. Section 5 proposes how this technology can be used in computer security with Sections 6 summarizing and proposing future work.

2 Background

E-commerce and marketing firms have taken advantage of profiling for years by collecting volumes of information on individuals. Such profiling is accomplished by aggregating information on individuals purchase history (online and offline), finance records, magazine sales, supermarket savings cards, surveys, and sweepstakes entries, just to name a few. This information is then cleaned, organized, and analyzed using a number of statistical and data mining techniques to create a "shopping" profile of that individual. These profiles can then be used to target ad campaigns, personalize a shopping experience, or make recommendations on additional products a user may find they "can't do without". Amazon (see http://www.amazon.com) is a perfect example of the effective conduct and implementation of this type of profiling with its tailored *"recommendation"* and *"customers who bought"* features in addition to their personalized e-mail campaigns that let users know when items you may be interested in or have looked at in the past go on sale.

Network traffic profiling, on the other hand, is an emerging area quickly filling up with commercial products that record and plot network usage by selected characteristics. These profiles, however, operate only at the level of network sessions and flows, while application level information, which is crucial in user behavioral profiling, is not used. Products like Qradar (see http://www.q1labs.com) and Mazu Profiler (see http://www.mazunetworks.com) consider TCP/IP traffic patterns as memory-less indicators of behavior. While interesting, we seek a more comprehensive profile, targeted towards user behavior, instead of straightforward host traffic graphing.

A final area that has seen great success in the field of behavioral modeling is business process modeling. Although primarily used as a tool to model work flows in an organization for the purposes of optimization of productivity and efficiency, much research has been done on how to model an individual user's actions and key characteristics. Data collection is traditionally done through the use of questionnaires, interviews, and direct observation, and the field provides a great deal of information and valuable lessons learned on what data is needed and how it can be best utilized to create an accurate profile of an individual. Our automated data collection techniques have the potential to advance the science of business process modeling by providing an easy and efficient way to test hypotheses and gather observations.

3 Approach

OBAMM is a new technique that uses information about a user's web browsing activities only, to accurately categorize what the primary interest areas are of that individual. We passively sniff network traffic to obtain browsing information, not instrumentation of the machine's browser application. Based on what we term *"reverse category lookup"*, OBAMM provides for very accurate categorization models to be built from fairly minimal user data. While fundamentally different in its implementation, the approach being proposed follows the same general methodology used in data mining (see [9]); collect data, process data, discover patterns, analyze patterns. The remainder of this section will briefly describe how each of these is accomplished using OBAMM.

3.1 User Data

With online users taking advantage of the Internet for everything from research, to hobbies, to online shopping, it would seem that all the information needed to describe a user is ready and available. While research in specific areas has been done to take advantage of users browsing activities in areas such as improving searches in peer-to-peer networks [10] and recommender systems for publications and retail [11, 12], little has been done to use this information to create a more complete user behavioral profile. The critical piece of information that provides the details needed for these types of applications comes in the form of the Uniform Resource Identifier (URI). The URI represents the global address of documents and resources that are present on the World Wide Web. The most common form of the URI is a web page address. When a user operates their web browser and requests a URI, or clicks a hyperlink on an existing page, a HTTP/GET request is generated. This GET request can be sniffed and captured, pulled from log files, or captured by agent devices installed on the user's machine. Timestamps can be added by recording the time that the GET request was captured, while the URI, the destination host, and assorted browser information are available directly from the request. By collecting the URIs that a user has visited, it is possible to have access to a large portion of the information that a given user has viewed in a period of time. By taking into account sites visited, frequency of visits, and duration of visit, it is straightforward to abstract information that provides a sketch of who an individual is.

3.2 Categorization Data

The key to OBAMM is the existence of a reverse lookup category data repository. This data comes in various forms, but for the purposes of this research, we are only interested in URIs that have been categorized in some hierarchical manner. An example of this type of hierarchical data can be seen in the Open Directory Project (ODP). ODP is an open content directory of World Wide Web links that has been constructed and categorized by humans. Web URIs based on similar content are grouped at a high level category while lower level sub-categories define varying levels of specificity for each site. For example, category information for http://www.dartmouth.edu would return the following category hierarchy:

Reference: Education: Colleges and Universities: North America: United States: New Hampshire: Dartmouth College

This information can be transformed into a directed graph G=(V,A) where V is the set vertices represented by the category description and E is the set of edges connecting each category. Representing the above category in this manner would yield the directed graph in Figure 1.

Figure 1: Graphical Representation of Category Information

As reverse categorization is done on additional URIs, the respective graph information can be added creating a graph representation of the users browsing activity. Edge weights are added to represent frequency while general category interest is represented by node weight. Building the graph in this manner provides an easy way to visualize a user profile as well as see data correlations that may have otherwise gone unnoticed. For example if dealing with the URIs http://espn.go.com and http://www.vivacheap.com it may not be intuitively obvious of any relation between these two sites. Storing the category information as a directed graph allows this correlation to be determined both visually (Figure 2) and mathematically using matrix algebra. Constructing the graph that represents this information illuminates the fact that both URIs share a common sports node. This may signify the beginning of a trend that can be used to begin to describe the individual. In addition, clustering techniques can be employed in order to allow the data to be viewed at varying levels of abstraction while pruning algorithms may be used to filter potential "noise".

A number of resources in addition to ODP provide commercial and open source options for gaining this type of URI categorized information (YellowPages.com, http://kc.forticare.com/). An important point that must be considered when utilizing multiple sources for this type of information is that they all must be normalized to the same format and category structure before being used for reverse category lookup.

Figure 2: Visual Correlation of Seemingly Disparate Data

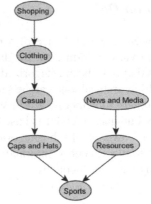

3.3 Pattern Discovery/Analysis

The actual format of a user profile is at the core of our technology. While a number of commercial tools exist today that claim to monitor and track user profiles and behaviors, most are only working at the TCP level and are doing little to identify or describe an individual based on what they are actually doing. When we consider behavioral profiles, three distinct orders of models can be described:

1. 0th order models: binary event recorded only, for instance: the list of websites that were visited.
2. 1st order models: frequencies, probability distributions, for instance: a Bernoulli style model indicating the likelihood that a site will be visited, based on a frequency count of previous visits.
3. 2nd order models: causality relations, time-dependencies, for instance: a hidden-Markov style model, with transitional probabilities between site accesses.

Most behavioral profiles fall in the 1st order model category, meaning they do not model important time-ordered sequences of events. For instance it is more likely that someone will visit their preferred shopping website first before going to others. Likewise, most people visit news and email websites, on a daily basis, in exactly the same order. These are important signatures of user behavior. In our work, we consider a user profile as a record of user browsing activity and can be described by the following key characteristics; destination, frequency, duration, and order. Destination represents where the user is actually browsing to. This information provides details on likes, dislikes, hobbies, interests, and other details relating to the individual. Frequency represents the number of individual times a destination is visited. This information combined with duration alludes to the importance of that data to the given individual. Order relates to the sequencing of the browsing activities from which it is possible to derive patterns of behavior about a given user. Although based solely on a users browsing activities, the aggregation of this information provides detailed attributes of the individual that can be used for the purposes of classification. By basing our profile on a combination of 0th, 1st, and 2nd order models, we have the ability to not only graph what the user is doing, but how they are doing it. Figure 3 is a graphical depiction of the categories of information describing a given user and the temporal interaction between them.

Figure 3: Temporal Aspects of Categorical User Data

4 Experimental Results

To test OBAMM, we chose to use ODP and a portion of the blacklist archive from URL Blacklist.com as our categorization dataset. At the time of the experiment, the DMOZ data set contained 4,830,584 URIs broken into sixteen main categories while the blacklist data contained 2,555,265 URIs split into 69 categories (all of which represented either primary or sub-categories of the DMOZ data). Normalization was done to the blacklist data in order to match the format and overall category structure of the DMOZ data set. The combined category archive was then hashed by URI. While it is understood that more optimal methods exist for the storage and retrieval of this type of information, for the purposes of this experiment, a has proved sufficient in regards to speed and accuracy.

Once all categorization data was collected and stored, approximately one week's worth of sample data was obtained from our local research lab, monitoring consenting individuals only. Data was preprocessed to include just the HTTP GET requests. The data was then broken out by source IP (where it is assumed that each unique source IP represents a unique user) and cleaned to remove adware and other automatically generated GET requests. The final data consisted of files each representing an individual user containing URIs of all the sites visited by that user. Data from each file was then run against the categorization hash in order to extract the category information re-

105

quired to construct a directed graph with category names representing nodes and link weights representing the number of times the user visited a site relating to that category. Graph information was stored in GraphML (see http://graphml.graphdrawing.org) file format so that it could be easily viewed, analyzed, and manipulated using graphical tools such as yED (see http://www.yworks.com). Figure 4 shows a subset of a user graph, representing the users browsing habits as they relate to the category *Computers*.

Figure 4: Subset of a directed graph representing a user's browsing habits as they relate to computers

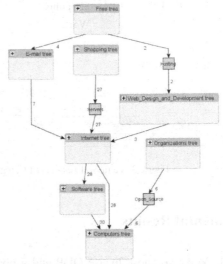

As can be seen in Figure 4, it is immediately apparent at a high level what type of activities the user was taking part in as well as the frequency of those activities. One of the keys to OBAMM, however, is the ability to view the users activity at varying levels of abstraction. Each of the larger boxes in the figure actually represents a grouping of two or more nodes in undirected subtrees of the principle node. Figure 5 is an example of how the *Shopping tree* node can be further expanded to display more detailed specifics about the user. From here, specific details can be obtained as to specifics about the individual, in this case in the form of shopping preferences.

Five individuals were broken out of the data set and categorized in this manner with approximately one week's worth of data for each user. Overall, 63% of the URIs were accurately categorized using only ODP and one category from the blacklist data set. While not a seemingly significant number at first glance, a number of factors are worth noting. First is that only one true categorization source was utilized (only one category from the black list data was used). Second is that because a hash was used, only exact matches were categorized. After closer examination of the user data, it is believed that we could increase the categorization percentage by 15-20% by utilizing a data structure that allows for fuzzy matching.

5 Security Applications

Using this methodology, we were able to quickly generate profiles summarizing the topics of interest to each user, how much time is spent on each topic, and how many times a topic or URI is visited.

Figure 5: Expansion of *Shopping tree* category node

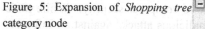
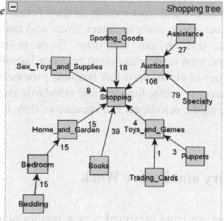

These profiles are very unique indicators of a persons online behavior, provide insight in to a number of key interest areas:

1. Malicious compromise. Hosts showing increased levels of activity at times of day, or volumes that are atypical for the profile on record. This could indicate a compromised machine exfiltrating sensitive information to an outside attacker.
2. Activity outside of normal work hours can be an indication of a malicious insider using the trusted environment to obtain and distribute sensitive documents.
3. Business process modeling. Unusually high levels non-work related browsing activity is an indication of less than optimal work efficiency by employees.
4. Out-of-band business processes. Two or more users in a network spend unusual amounts of time communicating with the same group of websites might indicate an ad-hoc business process that was created by users to circumvent an inefficiency in the organization. Not only can additional efficiency be gained by detecting and optimizing this step, it is also a security risk when an outside third-party is used for exchange of sensitive, mission critical information.

Although these security and efficiency implications are obvious, the potential impact of this intelligence is not. Many of today's successful organizations are held together by tightly guarded intellectual property that is at constant risk of compromise, either by insiders, outside attackers, or negligent behavior. Off-the-shelf network security solutions focus on packet or file monitoring for known-bad signature detection. So far little effort has been expended to detect users' behavior, which is the ultimate objective of our research. A direct correlation to this can be seen in the areas of business processes and trust models as they relate to computer security. While tools and techniques are in place to detect direct and malicious attacks against an information system, less obvious and often times un-intentional violations of business processes and trust models can often be more harmful. One of the issues with detecting these types of violations is that business processes tend to address the more general aspects of security rather than specific standards and protocols (i.e. not storing corporate data on public resources) and because of this, make detection difficult to impossible.

6 Summary and Future Work

In this paper we have presented a new methodology to conduct behavior modeling using pre-categorized data and emphasized how this data can be used in the field of computer security. While not a "silver bullet" in and of itself, our initial testing has demonstrated the huge potential for using this type of information. A network security solution that monitors email for IP leakage is easily circumvented by using a printer, or a USB memory stick. The detectable event, however, is therefore not the actual exfiltration, instead it is the process of the user obtaining the data, and the intent to illegitimately redistribute. Since this is a "behavior", not an "event", some heuristic or stochastic estimation needs to be applied to deduce intent. We strongly believe that creating models of user behavior is the first step in this process, and that future research will decide if behavior profiling is a usable, and morally justifiable way of limiting an organizations exposure to IP related risks.

A number of key factors still require additional research for this methodology to be even more effective. While we are currently storing temporal information as it relates to users and categories, little is being done with it at this time. Research into how temporal aspects of this information can be used to effectively describe an individual is still needed. In addition, clustering and pruning algorithms need to be implemented more effectively to ensure to information of value is not inadvertently lost or miscategorized. Lastly, research into how to handle both search engine queries and URIs that cannot be categorized using this technique is required for the system to be complete.

References

1. Brown B, Aaron M (2001) The politics of nature. In: Smith J (ed) The rise of modern genomics, 3rd edn. Wiley, New York.
2. Olmez A E (2006), Exploring the Culture-Performance Link in Heterogeneous Coalition Oganizations. PhD Thesis, George Mason University, Fairfax, VA.
3. Smith J, Jones M Jr, Houghton L et al(1999) Future of health insurance. N Engl J Med 965:325-329
4. South J, Blass B (2001) The future of modern genomics. Blackwell, London
5. Tantipathananandh C, Berger-Wolf T Y, Kempe D (2007) A framework for community identification in dynamic social networks. In: proceedings of the 13th international conference on knowledge discovery and data mining 717-726
6. V. Berk an G. Cybenko, "Process Query Systems", IEEE Computer, January 2007, p 62-71
7. Qradar, http://www.q1labs.com
8. Mazu Profiler, http://www.mazunetworks.com
9. Ian H. Witten and Eibe Frank, "Data Mining, Practical Machine Learning Tools and Techniques", 2nd edition, MK publishers 2005
10. Lu J, Callan J (2006) User Modeling for Full-Text Federated Search in Peer-to-Peer Networks. In: Annual ACM Conference on Research and Development in Information Retrieval Proceedings of the 29th annual international ACM SIGIR conference on Research and development in information retrieval 332-339
11. Quatse, Jesse and Najmi, Amir (2007) "Empirical Bayesian Targeting," Proceedings, WORLDCOMP'07, World Congress in Computer Science, Computer Engineering, and Applied Computing
12. Parsons, J., Ralph, P., & Gallagher K. (2004) Using viewing time to infer user preference in recommender systems. AAAI Workshop in Semantic Web Personalization, San Jose, California, July.

Mining for Social Processes in Intelligence Data Streams

Robert Savell[†] and George Cybenko[†]

[†][rsavell, gvc]@cs.dartmouth.edu, Thayer School of Engineering, Dartmouth College, Hanover NH

Abstract This work introduces a robust method for identifying and tracking clandestinely operating sub-nets in an active social network. The methodology is based on the Process Query System (PQS) previously applied to process mining in various physical contexts. Given a collection of process descriptions encoding personal and/or coordinated behavior of social entities, we parse a network's transactional stream for instances of active processes and assign process states to events and functional entities based on a projection of the evidence onto the process models. Our goal is not only to define the social network, but also to identify and track the dynamic states of functionally coherent sub-networks. We apply our methodology to a real world security task— mining a collection of simulated HUMINT and SIGINT intelligence data (the *Ali Baba* simulated intelligence data set)— and demonstrate superior results both in partitioning and contextualizing the social network.

Key words: Social Network Analysis, SNA, Process Query System, PQS.

1 Introduction

In this work, we introduce a process based framework for detection and tracking of social processes operating on social networks. While informed by role and structural analysis techniques found in traditional Social Network Analysis (SNA), this work seeks to expand the realm of SNA beyond the traditional stationary analyses of individual roles and community structure by focusing directly on the processes supporting the socio-temporal dynamics of active social networks. This process driven approach to Dynamic Social Network Analysis (DSNA) adopts the view that a network's organizational structure as well as the characteristic messaging patterns among nodes are a reflection of the interaction of a collection of distributed coherent social processes operating on the network.

This paper presents our methodology for applying the process based paradigm to the analysis of the sparse and noisy event streams which are common in the intelligence domain. In an application to the Ali Baba Scenario 1 data set (simulating activities of an embedded terrorist group), we demonstrate that a process based analysis produces superior results in partitioning the event stream, identifying active functional subnetworks, and extracting *process signatures* or state sequences produced by the coordinating processes of these sub-networks.

2 Background

Social Network Analysis has traditionally focused on stationary network descriptions to perform a variety of structural analyses such as identification of cohesive subgroups and the characterization and classification of individual roles and relationships in a social network [5, 6, 7]. Unfortunately, stationary analyses tend to be ill-suited to dynamic situations in which network organization and message patterns vary with time. For example, in real-world networks, functional roles of individuals or sub-networks are often temporally interleaved, with entities operating in one functional capacity at one moment and a different capacity with different sub-net relationships in the next. Interleaved behaviors may be difficult or impossible to distinguish in a stationary analysis. Stationary SNA is also limited in its ability to locate sparse transactional channels. These sparse channels often reflect key relationships, but are ignored in an aggregate traffic analysis. In addition, without a mechanism for tracking dynamics, a stationary analysis technique lacks a mechanism for analyzing the transactional streams generated by the social network.

2.1 Process Based Detection: The Process Query System (PQS)

To overcome the limitations of stationary SNA techniques, our previous work successfully applied process detection techniques to dynamic social network analysis. In [3], we demonstrate that an application of the Process Query System (PQS) produces superior results when tasked to extract coherent message threads from the Enron corpus.

The PQS system may be understood as a software system that allows users to interact with temporally indexed streams of data from multiple sources. Whereas traditional databases accept user queries in the form of constraints on field values of records in the database, PQS system queries take the form of a process model or description. Given a process query, PQS searches the data stream for evidence of the existence of processes consistent with the model. For an in-depth description of PQS, see Cybenko et. al [4, 2].

The PQS framework posits a collection of processes: $\{\mathcal{M}_1, \mathcal{M}_2, \ldots\}$, producing an interleaved stream of events: $\ldots, e_i, e_{i+1}, e_{i+2}, \ldots$ where event e_j occurs at time t_j with $t_j \leq t_{j+1}$. The goal in many applications is to solve the inverse problem. This may be formally defined as:

The Discrete Source Separation Problem(DSSP): Given a finite sequence of observed events: $e_{t_1}, e_{t_2}, \ldots, e_{t_n}$ and a collection of processes $\{\mathcal{M}_1, \mathcal{M}_2, \ldots\}$, determine:

1. The "best" assignment of events to process instances, namely

$$f : \{1, 2, \ldots, n\} \to \mathcal{N}^+ \times \mathcal{N}^+,$$

   where $f(i) = (j, k)$ is interpreted as meaning that event e_i was caused by the kth instance of process model j (the process detection problem);
2. The corresponding internal states and state sequences of the processes thus detected (the state estimation problem). Here $\mathcal{N}^+ = 1, 2, \ldots$ is the set of positive integers.

3 Methodology

In this application, we shall seek to identify discrete sources $\mathcal{M}_1, \ldots, \mathcal{M}_x$ by evidence of socio- and spatio-temporal coordination patterns. As in the general Discrete Source Separation Problem, each source is completely described by the subset of transactions associated with that source. We map each reported transactional event to an n-tuple (consistent with the evidence format of Ali Baba Scenario 1 Version 3) of the form:

$$e_d : \{(\text{date,location, named entities}) \times 3\}$$

A full solution of the DSSP produces a partition of the event stream, with each item of evidence e_d assigned to a particular instance $k_1, k_2, \ldots, k_r$ of a particular process j, so that process instance $\mathcal{M}_{(j,k)}$ is defined by the subset of events of the transactional stream $e_1, \ldots, e_n$:

$$\mathcal{M}_{(j,k)} = \{e_{(d,j,k)} : j \in 1 : x, k \in 1 : r, d \in 1 : n\}.$$

3.1 Hierarchical Coordination Processes

We define a generic model of a persistent coordinating process which captures the essence of the generic synchronization events encountered in an active functioning organization. This paradigmatic process is described by a simple Finite State Machine (FSM) as shown in Figure 1. The central portion of the figure describes the persistent *MEETING, PLANNING,* and *EXECUTION* states of an organization, group, or subgroup as it goes about its semi-autonomous functions in pursuit of a goal. Synchronization activities with superiors occurs via the *SYNCH UP* facility. Synchronization and control of subordinate group processes occurs via *SYNCH DOWN*. The coordination process is designed to be embedded in a hierarchical organizational description, with meetings arranged with associated subgroups at the *CONVERGENCE* node, and tasking and dispersal into subgroups organized at the *DIVERGENCE* node.

Viewing the organization as a hierarchical collection of weakly coupled coordination processes, the current application attempts to discover an optimal partition of the named entities into functional subgroups which maximizes the spatio-temporal and socio-temporal synchrony within the associated event streams. Many potential measures of synchrony are available. In general, we seek to maximize some measure of correspondence $\Omega(\{\mathcal{M}_j\})$ of named entity j to other members of a sub-net given the source definitions: $\mathcal{M}_j = f(\text{date,loc,}\{\text{entities}\}) \to \{e_{(d,j,k)}\})$.

In addition to the generic coordination process, certain task specific process descriptions— process descriptions which directly encode the details of a particular behavior— can also prove to be useful. In the current context, we implement a preliminary scan for *broadcast events*. These open transmissions introduce a preponderence of irrelevant correlations into the network structure, and therefore we

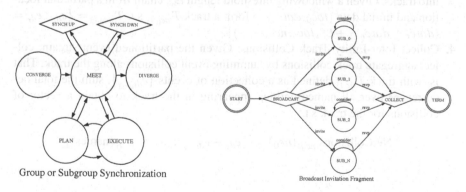

Group or Subgroup Synchronization

Broadcast Invitation Fragment

Fig. 1 FSM: Group Coordination Process.   **Fig. 2** FSM: Invitation Broadcast.

flag the associated events as noise in pre-processing. The Broadcast FSM is shown in Figure 2.

3.2 Organizational Process Discovery

In the analysts' database, we can expect to see sparse human and signal intelligence events which constitute a noisy signature of the activities of a potentially malevolent organization and its associates. Socio-temporal and spatio-temporal synchrony across active entities provides the most fundamental signature of the active organization. In our experiments, we implement a two stage analysis.

1. Entity Tracking: Segment the individual event streams according to spatiotemporal trajectories. Isolate low entropy events such as bursty broadcasts. Collect trajectory collision (co-occurence) statistics between named entities.
2. Structural Aggregation: Aggregate entities into structurally coherent units via a hierarchical clustering technique operating on the social network— with potentials and connectivity information acquired in Step 1.

3.2.1 Entity Tracking

Stage One of our method proceeds as follows:

1. Catalogue and remove broadcast events.
2. Infer a home location for each entity (based on event frequencies per location).
3. Partition the Event Stream: For each named entity ne_0, partition the event stream into tracks. Given a windowing threshold length L_w, chain from a particular location and initial date $(loc_0, date_0)$ to form a track $T_{(ne_0, i)}$ s.t. $T_{(ne_0, i)} = \{e_{d_r} : e_{d_r} = (date_r \leq date_{r-1} + L_w, loc_0, ne_0, -, -)\}$.
4. Collect Inter-Entity Track Collisions: Given the partitioned event stream, collect aggregate track collisions by summing event collisions along the track. That is, with track $T_{(ne,i)}$ defined as a collection of events $\{e_{d_{(ne,i)}}\}$, sum the total occurrences over all named entities appearing in the track to obtain a vector of collisions for the track s.t.:

$$\text{NetCollisions}_{(ne,i)}(ne_0) = \left| \{e_{d_x} \in e_{d_{(ne,i)}} : ne_0 \in \{e_{d_x}(\text{names})\} \right|$$

.

114

3.2.2 Structural Aggregation

In the next step, we partition the named entity set into relevant functional units. Since we are attempting to detect clandestine activities, we use a slightly different measure of functional coherence than we might in the case of trusting networks. As noted in [1], efforts of a malevolent process to conceal its actions tend paradoxically to increase susceptibility of the network to detection. The presence of sharp boundaries in the triangular structure of the network due to its tendency toward insularity makes the malevolent network more susceptible to detection via triangular clustering techniques.

We employ a relatively straightforward heuristic for partitioning the network into functional units consisting of an agglomerative clustering technique which, at each round, greedily selects the pair of clusters c_i and c_j which 1) meets a minimal criterion for direct connectivity and 2) maximizes the following estimate for triangular similarity, weighted by the inverse square of the cluster sizes:

$$S(c_i, c_j) = \frac{\mathbf{w}_i \cdot \mathbf{w}_j}{|\mathbf{w}_i + \mathbf{w}_i|} \cdot \frac{1}{(|c_i| + |c_j|)^2}$$

Here, vectors $\mathbf{w}_i$ and $\mathbf{w}_j$ correspond to the net collision scores collected in the previous section. Due to the sporadic nature of the traffic, we choose to approximate the triangular traffic measure, rather than measure it directly. The similarity score is designed to approximate the average triangle weight per n^2 potential edges in the merged cluster.

Results of the aggregation procedure are presented in the next section. A look ahead to Figure 8 clearly demonstrates the strength of the technique in this situation.

4 The Experiment

We applied our process analysis technique to the Ali Baba Scenario 1 database (designed by Jaworowski and Pavlak— Spring 2003)— a simulated collection of human and signal intelligence reports describing activities surrounding a terrorist cell based in Southern England. The Ali Baba raw data consists of a collection of approximately 1000 unclassified, fictitious word documents replicating intelligence reporting on the actions of suspected terrorist activity in Southern England. These documents are comprised of police reports, intelligence information reports (local HUMINT and detainee reports from abroad), FBIS (Foreign Broadcast Information System) reports, and local communications intercepts (Tac Reps).

4.1 The Story

The Ali Baba Scenario 1 takes place from April to September of 2003. The ground truth document for Scenario 1 describes a core cell of 10 associates. These cell members and their assignments are described in Figure 4. Figure 3 shows the network of first order links surrounding these core cell members.

It is expected that a skilled analyst— through direct examination of the intelligence collection— will be able to infer that members of the terrorist cell are planning to "bake a cake": (a euphemism for building a bomb) which will be targeted to blow up a water treatment facility near London, with the aim of contaminating the city's water supply with biological agents. In associated deception operations, it is the intention of Imad Abdul to take advantage of a close knit association of sympathizers and terrorists from other organizations to fill the air with fake chatter about numerous assassination plots and surveillance activities at several port facilities for actions against Israeli and US shipping interests.

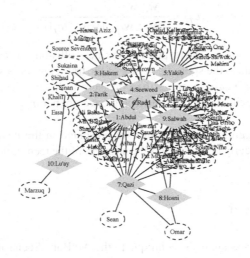

Fig. 3 Ground Truth: Ali Baba Scenario 1 Network.

5 Results

The ordered list in Figure 5 results from the application of a typical stationary social analysis technique based on social centrality— with node precedence determined by

1. Leader: Imad Abdul.	1. Associate: Phil Salwah.
2. Planner: Tarik Mashal.	2. Leader: Imad Abdul.
3. Recruiter: Yakib Abbaz.	3. Recruiter: Yakib Abbaz.
4. Security and Deception: Ramad Raed.	4. Planner: Tarik Mashal.
5. Computer hacker: Ali Hakem.	5. Demolitions: Quazi Aziz.
6. Financier: Salam Seaweed.	6. (Potential Recruit): Fawzan.
7. Demolitions: Quazi Aziz.	7. (Unknown) Alvaka.
8. Demolitions: Hosni Abdel.	8. (Unknown) Afia.
9. Associate: Phil Salwah.	9. (Unknown) Mazhar.
10. Associate: Lu'ay.	10. Financier: Salam Seaweed.

Fig. 4 Alibaba Scn1: Ground Truth.   **Fig. 5** Alibaba Scn1: SNA (Central Nodes).

the number of social triads associated with each node. As evidenced by the divergence of the list from ground truth, the stationary analysis successfully identifies several of the key players; however, there are significant omissions of key network members, as well as substitution of irrelevant entities. Most telling is the fact that the prominent network role is assigned to a peripheral associate (Salwah) due to his sociability outside the network. In this work, we successfully overcome the limitations of stationary SNA techniques by incorporating a sense of the process trajectories associated with the individual entities.

Figures 6 and 7 reflect the process signatures generated by several of the clusters and individuals. In each of these figures we project events associated with a sub-net or individual onto a simplified version of the generic process model. The resultant signatures consist of four event tracks. From lower to upper tracks, the figures indicate 1) events located at the entity's home location, 2) events assigned to away locations, 3) upward synchronization events, and 4) downward synchronization events. The X axis delineates the 225 days of the study. Plots in Figure 6 correspond to events produced by 1) The top 10 suspects in the Ali Baba ground truth, 2) The Saaud led decoy plot, and 3) the Irish army decoy plot. Hollow circles in the lower two tracks indicate spatio-temporal convergence of several individuals at a location (possibly a meeting). The second column of plots (Figure 7) correspond to the signatures of three of the core members of the sub-net.

Figure 8 demonstrates the efficacy of the methodology in extracting the relevant details of the malevolent sub-network. In the figure, nodes are labeled with the depth assigned to the node by the greedy clustering heuristic in the order of collapse into the final cluster. To maintain legibility of the plot, each of the nodes is connected to the sub-net body by its most significant edge (according to our clustering measure). Note the near exact correspondence with the expected expansion of the network space in accordance with the ground truth of AliBaba Scenario 1. As shown in Figure 8, ten of first twelve identified nodes or clusters is a member of the AliBaba network. In addition, the order of discovery respects the qualitative aspects of the Scenario 1 Ground Truth. Note that the peripheral *bridging* associate Salwah is now

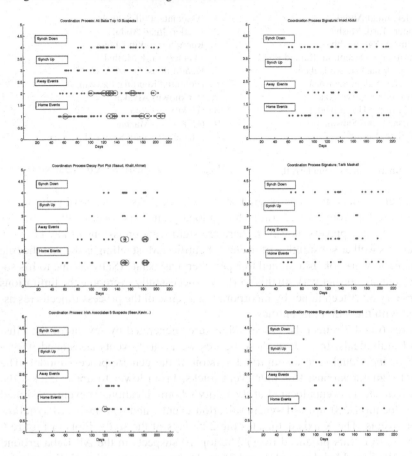

Fig. 6 Signatures: AliBaba and decoys.   **Fig. 7** Signatures: Abdul, Mashall, Seeweed.

in the proper position at the periphery of the core group. Also, decoy plots led by Khatib, Uvmyuzik and Sean are collected into their respective subnets and thinly connected to the group. Finally, we note the direct connection of Abdul to Ali Baba in an upward synchronization event.

In conclusion, we have demonstrated that the hierarchical organization of active processes can be extracted from the transactional patterns available in an intelligence data stream, without recourse to extensive parsing of informational content. Results of our hierarchical clustering are quite accurate relative to the ground truth of the Ali Baba Scenario 1 used in the study.

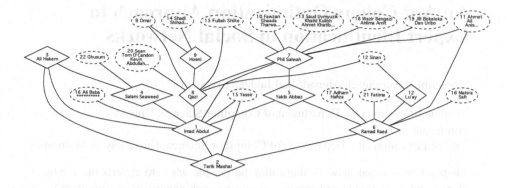

Fig. 8 Ali Baba neighborhood structure.

Acknowledgements This work is supported by DCI Postdoctoral Fellowship Grant HM1582-05-1-2033. Points of view in this document are those of the authors and do not necessarily represent the official position of the sponsoring agencies or the U.S. Government.

References

1. J. Baumes, M. Goldberg, M. Magdon-Ismail, and W. Wallace. On hidden groups in communication networks, 2005.
2. V. Berk, W. Chung, V. Crespi, G. Cybenko, R. Gray, D. Hernando, G. Jiang, H. Li, and Y. Sheng. Process Query Systems for surveillance and awareness. In *Proceedings of the Seventh World Multiconference on Systemics, Cybernetics and Informatics (SCI 2003)*, Orlando, Florida, July 2003.
3. W. Chung, R. Savell, J.-P. Schtt, and G. V. Cybenko. Identifying and tracking dynamic processes in social networks. In *Sensors, and Command, Control, Communications, and Intelligence (C3I) Technologies for Homeland Security and Homeland Defense III*, volume 6201 of *Proc. SPIE*, 2006.
4. G. Cybenko, V. H. Berk, V. Crespi, G. Jiang, and R. S. Gray. An overview of process query systems. In E. M. Carapezza, editor, *Sensors, and Command, Control, Communications, and Intelligence (C3I) Technologies for Homeland Security and Homeland Defense III*, volume 5403 of *Proc. SPIE*, pages 183–197, 2004.
5. W. de Nooy, AndrejMrvar, and V. Batagelj. *Exploratory Social Network Analysis with Pajek*. Cambridge University Press, Cambridge, UK, 2005.
6. T. A. Snijders. *Models and Methods in Social Network Analysis*. Cambridge University Press, Cambridge, UK, 2005.
7. S. Wasserman and K. Faust. *Social Network Analysis: Methods and Applications*. Cambridge University Press, Cambridge, UK, 1994.

An Ant Colony Optimization Approach to Expert Identification in Social Networks

Muhammad Aurangzeb Ahmad[1] and Jaideep Srivastava[2]

[1]mahmad@cs.umn.edu, Department of Computer Science, University of Minnesota
[2]srivasta@cs.umn.edu, Department of Computer Science, University of Minnesota

Abstract In a social network there may be people who are experts on a subject. Identifying such people and routing queries to such experts is an important problem. While the degree of separation between any node and an expert node may be small, assuming that social networks are small world networks, not all nodes may be willing to route the query because flooding the network with queries may result in the nodes becoming less likely to route queries in the future. Given this constraint and that there may be time constraints it is imperative to have an efficient way to identify experts in a network and route queries to these experts. In this paper we present an Ant Colony Optimization (ACO) based approach for expert identification and query routing in social networks. Also, even after one has identified the experts in the network, there may be new emerging topics for which there are not identifiable experts in the network. For such cases we extend the basic ACO model and introduce the notion of composibility of pheromones, where trails of different pheromones can be combined to for routing purposes.

1 Introduction

A social network comprises of many heterogeneous individuals. Social Networks can also act as knowledge markets and in such markets different people may have expertise in different fields or topics. Identifying all such experts is an important task. In a large network it may not be clear who the experts are. Additionally in a dynamic setting where the set of topics keeps on evolving it may not be feasible to keep track of the experts since the expertise of people can change and new topics may arise for which it may not be clear who the experts are. There are also issues related to scalability. In a small world network any node can be reached from any other node by a small number of hops and one could argue that one can just flood the network with queries to get an answer. However flooding the network with queries would certainly be counterproductive since people would be less likely to route queries if they are overwhelmed with requests to route queries. However if routing of queries is constrained by some criteria then the network traffic becomes manageable. Thus not all nodes may be willing to route queries in the network if they are routinely flooded with such queries. The aforementioned factors suggests a distributed approach to solving this problem.

120

Another factor to consider in this problem is that while it may be feasible to define a taxonomy for the topics *e.g.*, Michlmayr [7], in some settings it may not be desirable to do so as is the case of dynamically changing topics. In this paper we propose an approach which is inspired from the Ant Colony Optimization (ACO) metaheuristic which takes its inspiration from how social insects like ants forage for food. Whenever an ant comes across a food source it evaluates the quantity and the quality of the food source. The ant then goes back to the food source laying a trail of chemicals called **pheromones** in its path. The chemical trails gradually evaporates over time. As other ants come across this trail they also follow this path to the food source and also deposit pheromones. Thus the frequently used trails become stronger as compared to the less frequently used ones. As the food source becomes depleted, the ants look for alternative routes and the old trails becomes weak as well.

The problem of expert identification is applicable to settings where there is a large organization like a multi-national corporation with thousands of employees or the military with thousands of personals. Thus in a military scenario consider the case when a certain event like a terrorist attack warning may require finding people with certain expertise on a certain topic *e.g.*, Al-Qaeda money laundering in Yemen. There may not be an explicitly defined expert on that topic in the social network and given that there can be infinitely many such topics it is not feasible to keep track of such topics. An alternative is to pre-define topic hierarchies. However the problem with this approach is that there may be certain sub-topics that will fall into multiple categories and even some cases which do not even fall into any of the categories.

The approach outlined in this paper can be used in other distributed settings. The contributions of this paper are twofold. We describe a mechanism for expert identification for arbitrary topics. We run simulations which show that this approach fares better than the baseline. We also extend the traditional approach used in the ACO algorithms to cases where multiple pheromones can be combined for search in an ant network. The outline of the paper is as follows In section 2 we describe related work in numerous fields. In section 3 we give the details of the model and its variants. Experiments and results are described in section 4 where we compare the results with a k-random walker. An overview of the current work and future work is described in section 5.

2 Related Work

2.1 Ant Colony Optimization

The Ant Colony Optimization (ACO) metaheuristic was introduced by Marco Dorigo et al [4]. It has been applied to a wide range of combinatorial optimization problems like the Traveling Salesman problem, assignment problems, scheduling problems, vehicle routing problems etc[4]. AntHill [4] is a framework for the design, implementation and evaluation of ant algorithms in peer-to-peer networks.

Schelfthout et al. [8] explore the idea of using pheromones for coordination in distributed agent-oriented environments. For multiple keywords in the query their agents follow only one of the words but even with the simplification the hit rate is greatly increased. Schoonderwoerd et al. [9] use Ant Based Control (ABC) for routing in circuit switched networks. DiCaro and Dorigo's AntNet was devised for packet-switched networks. Elke Michlmayr et al. [6] proposed an algorithm *SemAnt* which, given frequent enough queries in a network can route queries efficiently given a pre-defined taxonomy. The current work builds upon SemAnt, however the main difference in their work the current work is that we do not assume a predefined taxonomy but rather allow for flexible taxonomies which can change over time.

2.2 Expert Identification

The problem of expert identification in social networks has been addressed before especially for e-mail based networks. Schwartz and Wood et al.[10] analyzed e-mail flows to identify groups of individual with common interests. The ContactFinder was a system that found the right person to forward a query using the text and addresses of messages on bulletin boards. Campell et al. [2] described a graph-based ranking approach that takes into account the content of email communication to determine experts in a network. Balog et al.[1] describe two strategies for expert identification. In the first method the expert is identified based on the documents that they are associated with. In the second approach they identify the documents associated with the topics and then identify experts associated with the topics.

2.3 Query Routing in Social Networks

A large amount of literature exists on routing in peer to peer networks which includes work in leveraging social networks for routing peer to peer networks. Here we describe some of the representative papers relevant to our work. For routing queries in distributed networks the most obvious techniques are based upon simple broadcasting. Histories of previous queries in the network have been used extensively to predict which nodes are likely to answer a query. Such techniques are considered to be reputation based. Thus Cohen et al. [3] note that nodes that are likely to answer a query are the ones which have answered similar queries in the past. Kleinberg et al. [5] describe Query Incentive Networks where users can pose queries in a network along with an incentive which can propagate across the network. They gave a game theoretic formulation of this problem and showed that a Nash Equilibrium exists for this game. Tempich et al. [11] describe an algorithm REMINDIN' that employs a similar idea wherein the nodes observe which nodes in the network have answered similar queries in the past and use this information to route queries. Similarly Tomiyasu et. al [12] describe a query routing mechanism for routing queries in a mobile cellphone network.

3 An ACO Model for Expert Identification

Here we describe a general overview of the ACO metaheuristic. Given a connected graph, a beginning node and a destination node who location is unknown, a node generates an 'ant' called the **Forward Ant** that tries to find the path to the destination node based on the previously laid trails of pheromones. A pheromone can be conceptualized as a marker which tells the ant which direction to tale. To avoid the coldstart problem the first c ants are just allowed to traverse the network like a k-random walker. For all the generated ants there is an associated **time to live** which specifies the maximum number of iterations that the ant should explore the network if the answer to the query is not found. If the destination node is not found then the ant is terminated. The forward ant also keeps track of the nodes that it has already visited. Alternatively if the destination node is found in iterations less than the time to live then the ant terminates at the destination node and creates a **backward ant** that retraces its path to the origin node. The backward ant then lays down the pheromone trail on this path. If Q is the length of the path and L_j denotes the distance from the origin to the node under consideration then the amount of pheromone laid on an edge j by an ant i is given by:

$$\Delta \tau_i^j = \begin{cases} Q/L_j, & \text{if } j \in \text{tour}(i) \\ 0, & \text{otherwise} \end{cases} \tag{1}$$

There is also an evaporation rate associated with the pheromones which describes the amount of pheromone that is decreased at each iteration. Given an evaporation rate ρ, the amount of pheromone τ is updated as follows:

$$\tau \leftarrow (1 - \rho) \cdot \tau \tag{2}$$

Scents in the trail are updated as new queries arrive. The query is routed to whichever node has the highest score. If j represents the indics of the nodes and U is the set of neighbors of the current node and F^Q is the set of nodes already visited by visited by the ant then j can be given as:

$$j = argmax_{u \in U \wedge u \notin (F^Q)} \tau_u \tag{3}$$

In the current setting, illustrated in figure 1, an ant is analogous to a query and the destination node is analogous to the expert. We consider two scenarios for topic distribution in the network. In the first case we assume that topics are predefined rigid categories while in the second case there can be potentially infinite number of pheromones and they can be composible. We now describe the basics elements of the ACO approach for our model. We represent the social network as a graph

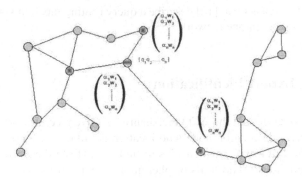

Fig. 1 The setting for the model. A social network where the query vector contains the query terms while the other vectors record the pheromone count for the nodes.

G with edge set E and the node set V where the nodes represent people and the edges represent ties between the people. There are also K nodes in the network which are also experts on certain topics. We vary the number of experts to see how well different approaches fare, however the case that we are most interested in is $|K| << |E|$. Each node i also maintains a vector v_i for all the topics or that it has encountered so far. The advantage of this approach is that no central representation of query scents is required. A query trail thus keeps track of the routes taken by the queries in the course of the history of the network.

In the second scenario, topics are not predefined but rather can be discovered 'on the fly.' An important reason for having non-rigid categories is that categories are likely to change over time, additionally some queries may not fall into well-defined categories. Thus given a query originating from a node n_o we determine its similarity to queries which have already been asked in the vicinity of the origin and route accordingly. The main difference between this approach and the traditional approach is that in this case we are not following a particular pheromone but rather a combination of different pheromones to determine the destination node. This also makes sense because in some cases there may not be a node in the network which has the exact answer to the query but the questioner may be interested in the next best answer. Although we are defining idea of similarity for combining pheromones for the domain of expert identification, this idea can be used in other domains as well. The pheromone in the second scenario can be represented as a k-dimensional vector which can encode more information about the topic as compared to the traditional view of pheromone. Thus given a query, at every iteration the distance between the query and pheromone on the node is calculated in order to determine which node should be selected for routing. We thus modify the rule for selecting which node to route to in the next iteration as follows.

$$j = argmax_{u \in U \wedge u \notin (FQ)} \left(\sum_{i=1}^{m} |d(\tau_i - \tau)| \right) \qquad (4)$$

In the traditional ACO model ants are terminated when they either find a route to the destination or if the time that the ant has lived is greater than time to live for the ant. In the current setting this formulation has to be modified so that we define a threshold for the similarity function such that if there is not an exact match for the query and the current time for the ant is equal to the time to live, then node in the path where the distance between the query and the pheromone on the node is least is returned as the destination node. To ensure that the ants do not get stuck in their neighborhood and thus miss out a potentially high quality unexplored trail nearby we use the idea of **Exploring Strategy** introduced in Ant Colony Systems [4]. At each iteration a random variable q is generated and based on the exploring strategy or the exploiting strategy is chosen. The strategy that was described previously *i.e.*, the strategy of choosing the nodes with the highest value for the pheromones is called the exploiting strategy. The exploring strategy however considers the potential of the unexplored nodes and visits those nodes with a certain probability even when there are nodes with a high pheromone value. The exploring strategy ensures that the forward ant not just considers the nodes with the highest pheromone values but also unexplored nodes before deciding which node to choose next. Thus associated with each unexplored neighbor node is a goodness value for that node which can be determined as:

$$p_j = \frac{|\tau_j - \tau|}{\sum_{u \in U \wedge u \notin (F^Q)} \left(\sum_{i=1}^{m} |d(\tau_i - \tau)| \right)} \tag{5}$$

One of the nodes is then selected based on the following rule.

$$GOTO_j = \begin{cases} 1, & \text{if } q \leq p_j \wedge j \in U \wedge j \notin S(F^Q) \\ 0, & \text{otherwise} \end{cases} \tag{6}$$

where q is a random number. As described previously not all the nodes may be willing to route the queries to their neighbors because of numerous reasons like bandwidth or unwillingness to forward messages which could be defined as being intrinsic. we define a motivation factor for routing for routing queries for all the nodes. Thus for every iteration and for every ant a random number is generated and if the random number is less than the motivation factor then the query is route to the neighboring node otherwise the query is not routed.

4 Experiments

We compare the performance of the ACO based algorithm to the k-random walker. This is analogous to the situation where one does not know who the experts are and thus one randomly explores the network. Since the network in question is a social network we are assuming that the topology of the network is that of a small world as borne our by the literature in the field [13]. Thus we consider a small world network with 1,000 nodes with 4 as the starting degree and a low evaporation rate for the pheromones *i.e.*, 0.0001. The time to live for an ant is roughly proportional to

Fig. 2 Hit Rate vs. Time to Live, when the number of experts and the motivation factor is kept constant. (Dark Line: ACO Algorithm, Lighter Colored Line: k-Random Walker)

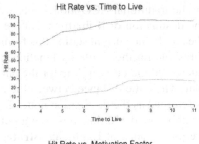

Fig. 3 Hit Rate vs. Motivation Factor, when the time to live and the number of experts are kept constant. (Dark Line: ACO Algorithm, Lighter Colored Line: k-Random Walker)

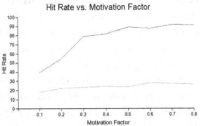

the hop distance from the origin node. For performance evaluation we use the same metrics as Michlmayr [7] Thus *Resource Usage* is the number of edges traversed for each query within a given period of time. *Hit rate* is the number of queries satisfied within a given period of time. *Efficency* is simply the resource usage divided by the hit rate. We performed exhaustive simulations for varying values of the size of the graph, the number of experts in the graph, the time to live and the motivating factor. Some of the representative results are described in detail below.

Figure 2,3,4 compares the average results for 2000 runs for the k-random walker against the ACO approach for different values of time to live, motivation factor and number of experts in the small world network. It is clear from Figure 2 that the ACO based approach performs much better than the k-random walker. In a social network the farther the query is from the origin node the less likely it is that the query will be answered, since the nodes which are farther away from the origin will be analogous to acquaintance of an acquaintance and thus likely to answer the query. Thus we will be interested in the hit rate for different values of time to live given that all the other factors are kept constant. Even for low values for TTL we get a fairly high hit rate. Figure 3 compares the performance of the two approaches when the motivation factor is varied and even in this case the ACO approach fares much better. Figure 3. Finally figure 4 compares the two approaches when the number of experts are varied. From the figure it is evident that ACO again outperforms the k-random walker and get a perfect hit rate *i.e.*, finds all the experts if five percent or more of the nodes are experts. In the composite case we considered the simplest cases, ranging from five to a hundred basic terms for queries and where the similarity can be defined for the terms. We note that similar results were obtained for experiments with composite pheromones where the ACO based algorithm performed better than the baseline.

126

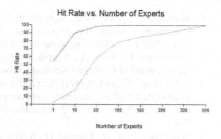

Fig. 4 Hit Rate vs. Number of Experts, when the motivation factor and the time to live is kept constant. (Dark Line: ACO Algorithm, Lighter Colored Line: k-Random Walker)

5 Conclusion and Future Work

In this paper we presented a model for expert identification and query routing in social networks based on the Ant Colony Optimization metaheuristic. The results were compared to a k-random walker as the baseline and the ACO based approached performed much better than the baseline for the general case and for composible pheromones also. The current model can be extended in a number of ways. The model described above assumes that the number of nodes in the network remain constant. In real world networks however the number of nodes does not remain constant as people can join, leave and even rejoin social network. Thus a possible line of research would be to consider the cases when new nodes are added to the network *i.e.*, addition of 'regular' nodes and the addition of expert nodes in the network. In the current model we assume that the search is conduced over keywords and keyword similarity can be used to describe topics which have not been encountered before. One way to extend the current work would be to use document similarity instead of keyword similarity. It should also be noted that recommendation systems in general suffer from the problem of cold start. One way to solve this problem would be to first identify experts in the network and use the experts for recommendation purposes. A scheme similar to the current one could be used for recommendation purposes.

References

1. Balog, K., Azzopardi, L., Rijke, M., Formal models for expert finding in enterprise corpora. SIGIR 2006: 43-50
2. Campbell. C.S., Magio, P.P.,Cozzi, A. Dom, Byron. (2003) Expertise Identification using email communications. CIKM 2003: 528-531.
3. Cohen. E., Fiat, A., and Kaplan, H. (2003) A Case for Associateive Peer to Peer Overlays. ACM SIGCOMM Computer Communications Review, 33(1):95-100, January 2003.
4. Dorigo, M., Blum, C., (2005) Ant Colony Optimization Theory: A Survey, Theoretical Computer Science 344 243-278.
5. Kleinberg, J. and Raghavan, P. Query Incentive Networks. (2005) In FOCS '05: 46th Annual IEEE Symposium on Foundations of Computer Science. Pittsburgh, PA, 132–141 2.

6. Michlmayr, E., Pany, Graf., S. Applying Ant-based Multi-Agent Systems to Query Routing in Distributed Environments, Proceedings of the 3rd IEEE Conference On Intelligent Systems (IEEE IS06), London, UK, September 2006

7. Michlmayr, E., Pany, A., Kappel, G., Using Taxonomies for Content-based Routing with Ants, Journal of Computer Networks, Elsevier, 2007

8. Schelfthout, K., and Holvoet, T., (2003) A Pheromone-Based Coordination Mechanism Applied to Peer-to-Peer. In Agents and peer-to-peer Computing, Second International Workshop (AP2PC 2003) volume 2872 of Lecture Notes in Computer Science, 71-76. Springner, July 2003.

9. Schoonderwoerd, R., Holland, O., Bruten, J., Rothkratz, L. (1996), "Ant-based load balancing in telecommunication networks", Adaptive Behaviour, Vol. 5 pp.169-207.

10. Schwartz, M.F., and Wood, D.C.M., (1993) Discovering shared interests using graph analysis. Commnications of the ACM, 36(8):78-89, 1993.

11. Tempich, C., Staab, S., and Wranik, A., Reminding: Semantic Query Routing in Peer-to-Peer Networks Based on Social Metaphors, Proc. 13th International World Wide Web Conf., pp. 640-649, 2004.

12. Tomiyasu, H., Maekawa, T., Hara, T., Nishio, S. Profile-based Query Routing in a Mobile Social Network. MDM 2006: 105

13. Watts, Duncan J.; Strogatz, Steven H. (June 1998). "Collective dynamics of 'small-world' networks". Nature 393: 440-442

Attribution-based Anomaly Detection: Trustworthiness in an Online Community

Shuyuan Mary Ho[†]

[†] smho@syr.edu, School of Information Studies, Syracuse University, Syracuse, New York

Abstract This paper conceptualizes human trustworthiness[1] as a key component for countering insider threats in an online community within the arena of corporate personnel security. Employees with access and authority have the most potential to cause damage to that information, to organizational reputation, or to the operational stability of the organization. The basic mechanisms of detecting changes in the trustworthiness of an individual who holds a key position in an organization resides in the observations of overt behavior – including communications behavior – over time. "Trustworthiness" is defined as the degree of correspondence between communicated intentions and behavioral outcomes that are observed over time [27], [25]. This is the degree to which the correspondence between the target's words and actions remain reliable, ethical and consistent, and any fluctuation does not exceed observer's expectations over time [10]. To be able to tell if the employee is trustworthy is thus determined by the subjective perceptions from individuals in his/her social network that have direct business functional connections, and thus the opportunity to repeatedly observe the correspondence between communications and behavior. The ability to correlate data-centric attributions, as observed changes in behavior from human perceptions; as analogous to "sensors" on the network, is the key to countering insider threats.

1 Introduction

According to the 2002 CSI/FBI Annual Computer Crime and Security Survey [24], insider misuse of authorized privileges or abuse of network access has caused great damage and loss to corporate internal information assets. Employees, as corporate assets, benefit an organization. But their inside knowledge of corporate resources can also threaten corporate security due to the possibility of improper disclosure of sensitive

---

[1] This paper comprises the framework and initial findings of my dissertation work on attribution-based behavioral anomaly detection.

information, whether unauthorized or authorized. Sensitive information that is compromised could even be maliciously used to counter-attack the organization.

This paper contains four major sections describing this study. Insider threat is identified in the *Problem Gaps*. The *Conceptualization* section discusses the possibility of predicting uncertain human behavior and defines the framework of attributing trustworthiness. The *Operationalization* section includes the rationale and challenging considerations for the methodological design, how the pilot was operationalized and initial findings on the mechanism of attributing trustworthiness. The *Conclusion* section justifies and summarizes this research in progress.

2 Problem Gaps

Since Robert Hanssen, a US counterintelligence agent, started spying and gave away highly classified national security documentary materials to KGB[2]/SVR[3] in Soviet Union / Russia in 1970, the case of a betrayal of trust by a trusted, high-ranked insider was established [7]. This case shows that not only the trust level of a key person with high-level security clearance could be altered, but that the danger s/he brings to corporate security is maximized as s/he knows what and where the critical corporate resources are. In another example, Johnson Pollard, with high-level security clearance, passed classified U.S. information to the Israelis, and was arrested in 1985 [2], [12], [20], [23]. Not only public sectors have suffered from the possibility of insider threats, but organizations in private sectors have also been subject to illegal (internal) transactions. For example, the Enron scandal uncovered in 2001 was one of these cases. With 21,000 employees spread throughout 40 countries, Enron's executives lied about its corporate profits and concealed its corporate debts. This is a case of how business executives with *high social power* can perpetrate a high-level scam and put the entire company at risk [1]. In another case, a former Boeing employee who knew the corporate internal network and file locations, was charged with a data theft in 2007 [4].

---

[2] KGB (transliterationof "КГБ") is the Russian abbreviation for Committee for State Security (Комитéт Государственной Безопáсности).

[3] SVR is the Russian abbreviation for Foreign Intelligence Service (Служба Внешней Разведки), which is Russia's primary external intelligence agency.

The phenomenon of insider threats is a social, human behavioral problem. C^4ISR^4 Joint Chiefs of Staff [3] believe that in addition to integrating technology, policy, management, and procedure, "human factors" have greatly threatened and caused vulnerability to the chains of defense [3:5]. In the *Insider Threat Study by CERT (2004-2005)*, US DoD[5], DHS[6], & Secret Service investigated various insider threat cases and discovered that embedded in a mesh of communications, a person given high social power but with insufficient trustworthiness can create a single point of trust failure [18]. Thus, "insider threats" as an organizational problem gap is defined as executives or someone with authorized access, high social power and holding a critical job position, who is capable of inflicting high impact damage including psychological, managerial, or physical level, within an organization. As Mitnick noted, bolstering against the weakest linkage - the human factor - becomes critical in the chain of security defense [21].

3 Conceptualization

The attribution-based anomaly detection on human trustworthiness is conceptualized in this section.

3.1 Predictability of Uncertain Human Behavior

Anomalous behavior can be caused by an internal trigger of personality disorder, or an external trigger of motivation. Prediction of the intended behavior is contingent on an individual's attitude, norm and perceived behavioral control. Although behavior can be measured in different dimensions, its predictive power does not always hold true with consistency and regularity.

While crime can be caused by "anomie" or "strain" behavior [6] due to an individual's peculiar way of thinking, societal change and culture also has an impact on criminal behavior [22]. Many theories, such as the sociological theory of deviance, differen-

---

[4] C^4ISR stands for Command, Control, Communications, Computers, Intelligence, Surveillance and Reconnaissance. It is an architecture framework used by the US DOD.

[5] US DOD stands for the US. Department of Defense.

[6] DHS stands for the Department of Homeland Security.

tial association theory, and social learning theory examine how criminal behavior interrelates with the social environment. The *control balance* theory states that societal norms and personal attachments provide a counter weight that helps protect against potentially deviant behavior [29], [16], [15], [28]. As such, the seriousness of deviant behavior may be impacted by the powerful counter control (isolation-based personal interests) of the delinquent [29]. While the seriousness may not be the key to control balance, it serves as the conjuncture to situational risks that would influence the control balance desirability [29:405]. Triggers in the surrounding environment could also play important roles in facilitating criminal actions [8], [9], [5]. As a result, situational factors may cause "strain" criminal behavior.

Regardless of the underlying cause, questionable human behavior can be detected through indirect perception of others. Three components work together to give clues to the questionable human behavior: the target, the observer and the situation [11], [14], [30].

3.2 Trustworthiness Framework

Trustworthiness of a person who has *social power* [17] and holds a key position is communicated, and can be perceived or attributed by his social connections (mainly peers, subordinates and associates). Figure 1 depicts the conceptual framework for attributing trustworthiness. It utilizes a multi-level analysis, including the mixed "lenses" of organizational and individual norms. Primary constructs, including social power, trustworthiness and communication, are represented in rectangular boxes. Relationships among constructs are represented by bi-directional arrows.

Figure 1: Conceptual Framework

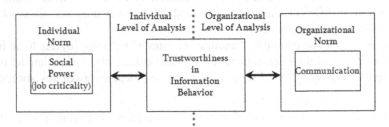

3.3 Definition of Trustworthiness

Julian Rotter [26] believes that "trust and trustworthiness are closely related." Rotter [27] defines trust as a generalized expectancy - held by an individual or a group - that the communications of another (individual or group) can be relied upon [25:652] regardless of situational specificity [27]. The term "expectancy" refers to a subjective probability - learned from previous experience - that an action will lead to a particular outcome [27]. Thus, one might say that the expectancy is in the "eye of the beholder." A generalized expectancy is one that holds broadly across a variety of situations. Interpersonal trust, defined by Rotter [25], [27], is "an expectancy held by an individual or a group that the word, promise, verbal or written statement of another individual or group can be relied upon." Hardin [10], on the other hand, differentiates trustworthiness from trust. He believes that trustworthiness is "a moralized account of trust" [10:28]. For example, a criminal (A) can trust a criminal-partner (B) to conduct a jointly orchestrated crime, and there is no moral or ethical notion involved. However, it might take much more complexity in decision-making for A to let B handle A's financial accounts because A may not find B trustworthy [10:29]. In this light, I define an employee's "trustworthiness" as the generalized expectancy, a subjective probability, toward a target employee's degree of correspondence between communicated intentions and behavioral outcomes that are observed and evaluated over time. In other words, the degree to which the relationship between the target's words and actions remain reliable, ethical and consistent, and any fluctuation in this relationship does not exceed the observer's expectations over time. To be able to tell if an employee is trustworthy is thus determined by the subjective perceptions of individuals in his/her social network who have direct functional connections, and thus the opportunity to repeatedly observe the correspondence between communication and behavior.

3.4 Attribution Theory

Attribution Theory intends to explain how people attribute (or assign) causes to another's behavior [13], [14]. It's a cognitive perception. Attribution theory dichotomizes behavioral causes to both internal and external. If the causes of behavior are attributed to the person, it is considered internal attribution. This causality of behavior is dispositional. If the causes of behavior are attributed to the situation, then it is considered to be external attribution. The basic observational "setting" contains three major variables: the observer, the target and the situation [11]. Perception can vary if the observations (or interpretation) are from different observers, or if the target being studied is different, or if the situation is different [14], [30:xvii].

4 Operationalization

Attribution theory is adopted in this experimetn to understand how people attribute (or assign) the causes of others' behaviors [13], [14]. The attribution of the target's (A's) behavior by observers (B's) is determined by B's judgment that A intentionally or un-intentionally [14] behaves in a way that is attributable to either external (situational) causality or internal (dispositional) causality. I set up a group of individuals (illustrated in Figure 2) as observers (B's) and collected daily perceptions about a target individual's (A's) behavior. A's behavior and the types of tasks involved were controlled, so that B's perceptions could be measured. A questionnaire was designed to collect any changes in B's judgments about A's behavior [14] that might be attributable to either external (situational) causality or internal (dispositional) causality. The principle of distinctiveness was also applied to A's behavior. In other words, behavioral change must be noticeable for others to perceive it. In order to eliminate the bias inherent in data collection, consensus among observers was obtained whenever possible. The consistency between the target's words and actions was evaluated by these observers [19] over time until a given set of tasks were completed.

Figure 2: Illustration of Experimental Situations Over Time. B_n perceive and assign meanings to A's behavioral consistency and level of trustworthiness.

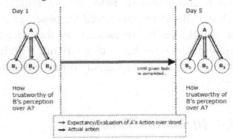

Consistency between a target's (A's) words and actions is an important indicator for properly measuring generalized expectancy through social communication to and from his/her peers. In this experiment, the consistency of the target's behavior was perceived by a group of observers. A couple of scenarios were schemed and classified in **Table 1**.

Table 1. Behavioral parameters determining trustworthiness

Target A's Behavior	Scenarios			
	1) Words ≠ Action	2) Words = Actions	3) Words > Actions	4) Words < Actions
Observer B's Perception	Questionable / Not Trustworthy	Trustworthy	Questionable / Not trustworthy	Not enough indication (positive vs. negative)

4.1 Experiment Design

This experiment was called the "Leader's Dilemma" Game[7]. It implemented the concept of a "virtual asynchronous contest," which was designed to recruit one real team and then share fictitious scores of three other teams (Figure 3). The "Leader's Dilemma" game was launched on Syracuse University's WebCT[8], and lasted for 5 consecutive days.

The bright blue area is the actual team recruited - the gray areas represent fictitious teams. Participants were recruited (n=5) for the pilot experiment. One participant was appointed as the team-leader (A). The remaining participants were team players ($B_{n=4}$). The experimenter (M) monitored the game play and collected participant observations. M's role in this controlled environment was a corporate executive, setting up the game rules and emphasizing the corporate interests. The game-master (G) was in an outside-authoritative position (representing market competition). G had insider knowledge of the competition and had the power to award the wining team. G's role was to inject "bait" (in the form of money tokens) during a point in the game, to entrap A if possible. While A was being entrapped, B_n perceived A's behavior. The perceptions from B_n were collected in various ways: participant observation, daily survey, archived online communication messages, and semi-structured interviews.

---

[7] This study is approved under IRB protocol #07-276.

[8] WebCT is a Learning Management System hosted by School of Information Studies, Syracuse University. Please go to http://ischool.syr.edu/learn/ for more information about WebCT.

Figure 3: Experiment Control Room Design   Figure 4: Logic for Virtually Controlled Contest

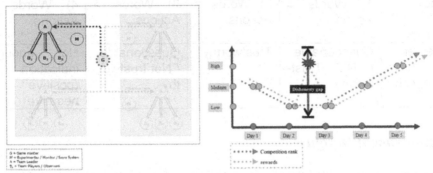

This virtual contest reached its climax when a "dishonesty gap" was forcefully created by feeding the bait solely to the team leader (Figure 4).

4.2 Pilot Experimentation

This pilot experimentation simulated the real world by partially disclosing the award system to the "winning" team, but hiding additional incentives from the participants.

This pilot experimentation reported that the game-master successfully manipulated and persuaded the team-leader (the target) to take the bait - in addition to what was awarded to the team in each day's game. The target faced a serious ethical dilemma and moral struggle in deciding how to distribute the team-based award and incentives to individuals once the monetary value was emphasized. The swift trust from the team players to the target was caused by the leadership halo effect - until Day 4. At this point, the target showed signs of dishonesty on a couple of occasions. In addition to the monetary bait injected to the target, the game-master used a negative team evaluation appraisal strategy on the target to stir up his disgruntled feelings about his team. However, the game-master showed understanding to the human weaknesses. He won the target's trust.

4.3 Initial Findings of Attributing Trustworthiness Mechanism

The participants did not know about the underlying deal between the game-master and the team-leader – which occurred in the background. Since insufficient evidence existed regarding the target (as **Table** 1), the resulting perceptions depended on whether the target's behavior was generally reliable or ethical, and the outcome itself. The perception of the target's behavior was positive from Day 1 through Day 4 during the experiment. As a result, the observers' attribution of the target was seen as being trustworthy. But, the outcome of the target's behavior was negative on Day 5, and the level of the target's trustworthiness dropped with regards to the observers' attribution (as Table 1). While participants found it difficult to discern the tone of conversations in the virtual room, they were able to attribute different personality traits to each identity.

Moreover, the participants were able to sense that the target became immediately defensive when an innocent question about the award distribution was asked. Some participants were able to attribute the discrepancy in the words spoken by the target. The target showed some inconsistency in what was said (words) versus what was done (behavior). According to the scheme laid out in Table 1, it indicated that the trustworthiness of this target was then perceived as questionable (not trustworthy). However, in a couple of situations where the target had confessed about some mistakes he had made in submitting the answers in the game (this target had already warned the observers), the subjective expectancy from his/her peers might reflect that this employee may be considered trustworthy since his/her words were consistent with his/her actions.

The attribution of trustworthiness was context-specific. Most of the attributions were centered on whether the target was responsible and accountable for his/her leadership. Some participants would base their judgment of the target's trustworthiness on how much others have trusted the target. In retrospect, their judgments followed the smoothness of the social interactions in the group. The same perceptions were applied by the leader in the evaluation of the team members. The leader would mostly look at the subordination, coordination, argumentativeness and submissiveness of attitude. An interesting phenomenon was noted; if a team player (as the observer) was too gullibly trusting[9] of others, it was harder for him/her to doubt the trustworthiness of others.

As a result of the relatively short duration of this experiment, the target was not too attached to the societal or group norms. The weaker bond to this group provided a stronger weight that enhanced the potential for deviant behavior. As such, the seriousness of deviant behavior was impacted by the powerful counter control (isolation-based personal interests) of the delinquent [29].

---

[9] Trusting refers to trustfulness.

5 Conclusion

In traditional information security strategy, in order to effectively quarantine polymorphic virus codes, it has been necessary to study how codes change. Likewise, we may be able to detect suspicious behavior and counter insider threats by studying how human behavior changes and how ones' trustworthiness is altered through indirect perception of others within an organization. Human's perception is not fully reliable due to the fact that not all information is made transparent to the perceivers. Humans attribute their perception of people's trustworthiness based on limited social interactions with a target. Most of the attributions are context-specific and are mixed with the judgment of the target's capability of holding responsibility and accountability for achieving external goals. Basic struggles of personal gain, selfishness and greediness remain - not only in physical environment - but in online community. The ethical values and moral standards are vaguely defined by the society and therefore vaguely adapted by individuals.

This study is in the early stage of developing an attribution-based behavioral anomaly detection model. It attempts to provide basic mechanisms, clues, and early warning signs to investigate and detect fluctuations in personnel trustworthiness; however it is important to note that these by no means provide full assurance to predict or convict crimes. While these basic mechanisms can be adopted for behavioral modeling of early warning systems, human intervention is still necessary in this loop. A behavioral anomaly detection toolkit could be produced and utilized to evaluate human trustworthiness as part of a larger corporate security preventive control system for the purpose of countering insider threats.

Acknowledgements I thank Jeffrey M. Stanton, my advisor, for his constant support and insight. I thank Conrad Metcalfe for his helpful comments and editing assistance. I also wish to thank Chandrasekhar Sridhar for his help in collecting pilot data.

References

1. BBC News Online. (2002). "Enron Scandal at a glance," [BBC News Online]. Obtained on August 22, 2006 from http://news.bbc.co.uk/1/hi/business/1780075.stm.
2. Benkoil, D. (1998, October) "An Unrepentant Spy: Jonathan Pollard Serving a Life Sentence," [ABCNEWS.com], October 25, 1998.

3. *C⁴ISR* Joint Chiefs of Staff (2000). *Information Assurance Through Defense In Depth*. Washington D. C., February 2000.
4. Carr, J. (2007). *Former Boeing Employee charged in data theft*. SC Magazine. Released on July 12, 2007. Obtained from http://www.scmagazine.com/us/news/article/670671/ex-boeing-employee-charged-data-theft.
5. Cohen, L.E. and Felson, M. (1979). Social Change and Crime Rate Trends: A Routine Activity Approach. *American Sociological Review*, 44(4), (Aug., 1979), 588-608.
6. Durkheim, E. (1897). *Suicide: A Study in Sociology*. Trans. John A. Spaulding and George Simpson. New York: The Free Press, 1951.
7. FBI National Press Office. (2001). *Federal Beaureu of Investigation Story: Robert Philip Hanssen Espionage Case*. Released on Feb 20, 2001. Obtained from http://www.fbi.gov/libref/historic/famcases/hanssen/hanssen.htm.
8. Felson, M. (1987). Routine Activities and Crime Prevention in the Developing-Metropolis. *Criminology*, 25(4), (Nov. 1987), 911.
9. Felson, M. and Cohen, L.E. (1980). Human Ecology and Crime: A Routine Activity Approach. *Human Ecology*, 8(4), (Dec. 1980), 389-406.
10. Hardin, R. (1996). Trustworthiness. *Ethics*, Vol. 107, No. 1. (Oct., 1996), pp. 26-42.
11. Hardin, R. (2003). Gaming trust. In E. Ostrom & J. Walker (Eds.), *Trust and reciprocity: Interdisciplinary lessons from experimental research* (pp. 80-101). New York: Russell Sage Foundation.
12. Haydon, M.V. (1999). "The Insider Threat to U.S. Government Information Systems", National Security Telecommunications and Information Systems Security Committee (NSTISSAM) INFOSEC 1-99, July 1999. http://www.nstissc.gov/Assets/pdf/NSTISSAM_INFOSEC1-99.pdf.
13. Heider, F. (1944). Social perception and phenomenal causality. *Psychological Review*, 51, 358-374.
14. Heider, F. (1958). The psychology of interpersonal relations. New York: John Wiley & Sons.
15. Hirschi, T. (1969). *Causes of Delinquency*. Beverly Hills, CA: University of California Press.
16. Hirschi, T. & Gottfredson, M.R. (1986). The Distinction Between Crime and Criminality. In *Critique and Explanation: Essays in Honor of Gwynne Nettler*, edited by T. Hartnagel & R. Silverman (pp. 44-69). NJ: Transaction.
17. Ho, S.M. (2008). *Towards a Deeper Understanding of Personnel Anomaly Detection*. Encyclopedia of Cyber Warfare and Cyber Terrorism, 2008 IGI Global Publications, Hershey, PA.
18. Keeney, M., Kowalski, E., Cappelli, D., Moore, A., Shimeall, T., and Rogers, S. (2005). *"Insider Threat Study: Computer System Sabotage in Critical Infrastructure Sectors."* National Threat Assessment Center, U.S. Secret Service, and CERT® Coordination Center/Software Engineering Institute, Carnegie Mellon, May 2005, pp.21-34. Obtained from http://www.cert.org/archive/pdf/insidercross051105.pdf on April 10, 2007.
19. Kelley, H.H. (1973). The Process of Causal Attribution, *American Psychologist*, Feb 1973, 107-128. Obtained from http://faculty.babson.edu/krollag/org_site/soc_psych/kelly_attrib.html on July 5th, 2007.
20. Lamar, Jr. J.V. (1986). Two Not-So-Perfect Spies; Ronald Pelton is Convicted of Espionage as Jonathan Pollard Pleads Guilty. *Time*, 16 June 1986.
21. Mitnick, K.D. and Simon, W.L. (2002). *The Art of Deception: Controlling the Human Element of Security*. Indianapolis, Indiana: Wiley.

22. O'Connor, T. (2007). <u>An Outline of Strain Theory;</u> adapted from, T. O'Connor, Varieties of Strain Theory. Retrieved on January 05, 2007 from http://www.homestead.com/rouncefield/files/a_soc_dev_19.htm.
23. Park, J.S. and Ho, S.M. (2004). Composite Role-based Monitoring (CRBM) for Countering Insider Threats. *Proceedings of Second Symposium on Intelligence and Security Informatics* (ISI), Tucson, Arizona, June 2004.
24. Power, R. (2002). *CSI/FBI Computer Crime and Security Survey.* Computer Security Issues & Trends, 2002.
25. Rotter, J.B. (1967). A new scale for the measurement of interpersonal trust. *Journal of Personality*, 35 (4), 651–665.
26. Rotter, J.B. and Stein, D.K. (1971). Public Attitudes Toward the Trustworthiness, Competence, and Altruism of Twenty Selected Occupations. *Journal of Applied Social Psychology*, Dec 1971, 1(4), 334–343.
27. Rotter, J.B. (1980). Interpersonal Trust, Trustworthiness, and Gullibility. *American Psychologist*, Jan 1980, 35(1), 1–7.
28. Sykes, G.M. and Matza, D. (1957). Techniques of Neutralization: A Theory of Delinquency. *American Sociological Review*, 22(6), (Dec., 1957), 664-670.
29. Tittle, C.R. (2004). Refining Control Balance Theory. *Theoretical Criminology*, 8(4), November, 2004, 395-428.
30. Weiner, B. (2006). Social Motivation, Justice, and the Moral Emotions: An Attributional Approach. Lawrence Erlbaum Associates, inc., Mahwah, New Jersey.

Particle Swarm Social Model for Group Social Learning in Adaptive Environment

Xiaohui Cui[1], Laura L. Pullum[2], Jim Treadwell[1], Robert M. Patton[1], and Thomas E. Potok[1]

[1]Computational Sciences and Engineering Division, Oak Ridge National Laboratory, Oak Ridge, TN 37831
[2]Lockheed Martin Corporation, St. Paul, MN 55164-0525

Abstract This report presents a study of integrating particle swarm algorithm, social knowledge adaptation and multi-agent approaches for modeling the social learning of self-organized groups and their collective searching behavior in an adaptive environment. The objective of this research is to apply the particle swarm metaphor as a model of social learning for a dynamic environment. The research provides a platform for understanding and insights into knowledge discovery and strategic search in human self-organized social groups, such as human communities.

1 Introduction

The notion of social learning has been used with many different meanings to refer to processes of learning and change of individuals and social systems. In this research, social learning refers to the process in which agents learn new knowledge and increase their capability by interacting with other agents directly or indirectly. Research of some social species indicates that these social species have a kind of social learning capacity to use knowledge provided by other group members to help the whole group quickly respond and adapt to a dynamic environment. Swarm Intelligence is the research field that attempts to design computational algorithms or distributed problem-solving devices inspired by the collective social learning behaviors of these social species. Particle swarm algorithm [10] is a type of Swarm Intelligence and was originally developed by Eberhart and Kennedy in 1995. Since 2004, researchers have successfully applied the particle swarm model in the simulation of the social behavior of animals [7] and strategic adaptation in organizations [2]. However, in terms of self-organized group social learning and collective strategy searching behavior for dynamically changing environments, there does not appear to be any mature or widely used methodology.

In this research, a modified particle swarm algorithm model, PArticle Swarm Social (PASS) model, is used to model the self-organized group's social learning

141

and collective searching behavior in a dynamically changing environment. Different from the randomly changing environment model used in many research efforts, a new adaptive environment model, which adaptively reacts to the agent's collective searching behaviors, is proposed. An agent based simulation is implemented for investigating the factors that affect the global performance of the whole social community through social learning. The objective of this research is to apply the particle swarm metaphor as a model of human social group social learning for the adaptive environment and to provide insight and understanding of social group knowledge discovery and strategic search in a changing environment.

This paper is organized as follows: Section 2 provides an introduction to the canonical particle swarm optimization algorithm. Section 3 describes the PASS model, which covers the individual social learning behavior, the adaptive environment and a social network model for social learning behavior. Experiment results will be discussed in section 4. Section 5 describes a verification and validation approach for social network models. The conclusion is presented in Section 6.

2 Particle Swarm Algorithm

The particle swarm algorithm was inspired by the social behavior of bird flocks and social interactions of human society [10]. In the particle swarm algorithm, birds in a flock or individuals in human society are symbolically represented as particles. These particles can be considered as simple agents "flying" through a high dimensional problem space searching for a high fitness value solution or strategy. It is the particle's personal experience combined with its neighbors' experience that influences the movement of each particle through a problem space. Mathematically, given a multi-dimensional problem space, the i_{th} particle changes its velocity and location according to the following equations:

$$v_{id} = w \times (v_{id} + c_1 \times rand_1 \times (p_{id} - x_{id}) + c_2 \times rand_2 \times (p_{gd} - x_{id})) \qquad (1)$$

$$x_{id} = x_{id} + v_{id} \qquad (2)$$

where, p_id is the location where the particle experiences the best fitness value; p_gd is the location of the highest best fitness value current found in the whole population; x_id is the particle's current location; c_1 and c_2 are two positive constants; d is the number of dimensions of the problem space; $rand_1$ and $rand_2$ are random values in the range of (0,1). w is called the constriction coefficient [8]. Each particle will update the best fitness values p_id and p_gd at each generation based on Eq.3, where the symbol f denotes the fitness function; $P_i(t)$ denotes the best fitness coordination; and t denotes the generation step.

$$f(P_i(t+1)) = \begin{cases} f(P_i(t)), & \text{if } f(X_i(t+1)) \leq f(P_i(t)) \\ f(X_i(t+1)), & \text{if } f(X_i(t+1)) > f(P_i(t)) \end{cases} \qquad (3)$$

The P_{id} and the coordinate fitness values $f(P_{id})$ can be considered as each individual particle's experience or knowledge. The P_{gd} and the coordinate fitness values $f(P_{gd})$ can be considered as the best knowledge that an individual can acquire from its neighbors through interaction. The social learning behavior in the particle swarm algorithm model is mathematically represented as the combination of an individual particle's experience and the best experience it acquired from neighbors for generating the new moving action. In the following section, we propose a modified particle swarm social (PASS) model for modeling the self-organized group social learning.

3 Particle Swarm Social Model

The PASS model describes individuals that are affiliated with different groups, seeking the highest fitness (profit) value solution or strategy in a changing environment. Individuals use the information provided by other individuals to enhance their capability in finding the highest fitness value solutions. The solution landscape will dynamically change as the group individuals search for the highest profit strategy configuration. This demands that the groups not only find a highly profitable solution in a short time, but also track the trajectory of the profitable solution in the dynamic environment. The group members do not have any prior knowledge about the profit landscape. Following two sections will describe the two elements of the PASS model: the adaptive environment and the individual social learning behavior.

3.1 Adaptive Environment

In the PASS model, the change patterns of the environment will be influenced by the collective behaviors of the social groups when these collective behaviors are effective enough to alter the environment. We define this kind of environment as an adaptive environment. To simulate the movement of the solutions, a test function, DF1, proposed by Morrison and De Jong [12], is used to construct the solution fitness value landscape. DF1 function has been widely used as the generator of dynamic test environments [1, 13]. For a two dimensional space, the fitness value evaluation function in DF1 is defined in Eq. 4. Where N denotes the number of peaks in the environment. The (x_i, y_i) represents each cone's location. R_i and H_i represent the cone's height and slope. The movement of the problem solutions and the dynamic change of the fitness value of different solutions are simulated with the movement of the cones and the change of the height of the cone-shaped peaks. It is controlled through the logic function eq. 5. Where A is a constant and Y_i is the value at the time-step i. The Y value produced on each time-step will be used to control the changing step sizes of the dynamic environment.

$$f(X,Y) = MAX[H_i - R_i \times \sqrt{(X - x_i)^2 + (Y - y_i)^2}]; \; (i = 1, ...N) \qquad (4)$$

143

$$Y_i = A \times Y_{i-1} \times (1 - Y_{i-1}) \tag{5}$$

In real-world applications, the evaluated fitness value cannot always be calculated precisely. Most of the time, the fitness value will be polluted by some degree of noise. To simulate this kind of noise pollution in the fitness evaluation, a noise polluted fitness value function [13] can be represented in eq.6. where η illustrate the noise and is a Gaussian distributed random variable with zero mean and variance σ^2. Another dynamic mechanism of the fitness landscape is the fitness value will gradually decrease with an increasing number of the searching group members that adopt similar solution. The eq. 7 represents this fitness value decrease. Where f is the landscape fitness value of strategic configuration (x, y) at the iteration i. N denotes the number of group member that adopts similar strategic configurations.

$$f^n(x) = f(x) \times (1 + \eta); \quad \eta \sim N(0, \sigma^2) \tag{6}$$

$$f_i(x, y) = f_{i-1}(x, y) \times \left(\frac{1}{e^{(N-1)}} \right) \tag{7}$$

3.2 The individual social learning behavior

Social learning behavior occurs when individual can observe or interact with other individuals. An individual will combine his individual experience and the information provided by other experienced individuals to improve its search capability. The particle swarm algorithm is used to control the individual's social learning behavior in the dynamical profit space. According to the eq.3, a particle's knowledge will not be updated until the particle encounters a new vector location with a higher fitness value than the value currently stored in its memory. However, in the dynamic environment, the fitness value of each point in the profit landscape may change over time. The problem solution with the highest fitness value ever found by a specific particle may not have the highest fitness value after several iterations. It requires the particle to learn new knowledge whenever the environment changes. However, the traditional particle swarm algorithm lacks a knowledge updating mechanism to monitor the change of the environment and renew the particles' memory when the environment has changed. As a result, the particle continually uses outdated experience/knowledge to direct its search, which inhibits the particle from following the movement of the current optimal solution and eventually, causes the particle to be easily trapped in the region of the former optimal solution. In the PASS model, a distributed adaptive particle swarm algorithm approach [9] is used to enable each particle to automatically detect change in the environment and use social learning to update its knowledge. Each particle will compare the fitness value of its current location with that of its previous location. If the current fitness value doesn't have any improvement compared to the previous value, the particle will use Eq.8 for the fitness value update. Eq.8 is slightly different than the traditional fitness value update function provided in Eq.3.

$$f(P_i(t+1)) = \begin{cases} f(p_i(t)) \times \rho, & \text{if } f(X_i(t+1)) \leq f(P_i(t)) \times \rho \\ f(X_i(t+1)), & \text{if } f(X_i(t+1)) > f(P_i(t)) \times \rho \end{cases} \qquad (8)$$

In Eq. 8, a new notion, the evaporation constant ρ, is introduced. ρ has a value between 0 and 1. The personal fitness value that is stored in each particle's memory and the global fitness value of the particle swarm will gradually evaporate (decrease) at the rate of the evaporation constant over time. If the particle continuously fails to improve its current fitness value by using its previous individual and social experience, the particle's personal best fitness value $f(p_{id})$ as well as the global best fitness value $f(p_{gd})$ will gradually decrease. Eventually, the $f(p_{id})$ and $f(p_{gd})$ value will be lower than the fitness value of the particle's current location and the best fitness value will be replaced. Although all particles have the same evaporation constant ρ, each particle's updating frequency may not be the same. The updating frequency depends on the particle's previous best fitness value and the current fitness value $f(X)$ that the particle acquired. The particle will update its best fitness value more frequently when the previous best fitness value is lower and the $f(X)$ is higher. Usually the new environment (after changing) is closely related to the previous environment from which it evolved. It would be beneficial to use the existing knowledge/experience about the previous landscape space to help particle searching for the new optimal. The Eq.8 enables each particle to self-adapt to the changing environment.

4 Social Learning Simulation Experiment and Results

The implementations of the PASS model and the adaptive environment simulations are carried out in the NetLogo agent modeling environment [15]. Each agent in the NetLogo environment represents one particle in the model. Initially, there are 400 agents randomly distributed in an environment that consists of a 100x100 rectangular 2D grid. The grid represents all the possible strategic configurations or solutions the agents may adopt for their profit. A dynamic profit landscape is generated as discussed in section 3.1 and mirrored on the 2D grid. The initial status of the 2D grid is shown in Fig.1(a). Eight white circuits represent the fitness (profit) values of solutions. The brighter the white circuit, the higher the fitness (profit) value is. These white circuits will dynamically move in the grid base on the Eq. 4 and the fitness values (brightness of the circuits) are dynamically changed base on Eq. 5 and Eq. 6. The agents are represented as the color dots in the grid. Different colors indicate different groups. The social learning of each individual is represented as the highest fitness value and location broadcast within the group. The searching behavior for finding highly profitable solution is represented as the movement of agent in the 2D grid. The movement of each agent is controlled by Eq. 1 and Eq. 2, in which c_1 and c_2 are set to 1.49, V_{max} is set to 5 and the w value is set to 0.72 as recommended in the canonical particle swarm algorithm [12]. It is assumed that agents belonging to the same group can exchange information without any restriction. But the information exchanged between different groups will be delayed for a pre-defined number of time-steps and some noise will be added to pollute the value of the information to reduce the information's accuracy. The delayed time-step for information exchange between agent groups is pre-set as 20 time-steps. There is a 20% possibility that the information, including the location of the best fitness value and the fitness value itself, is incorrect.

Fig. 1 The distribution of (a) initial environment, (b) scenario a: 1 group, 400 agents, (c) scenario b: 20 groups, 20 agents per group

In this experiment, we first investigated the change of a particle's social learning performance when the social group structure changed from a single group to multiple groups. The performance evaluation can be generated via computing the average fitness value of all individuals generated in the whole search period. Two different agent group structure scenarios, scenario a and scenario b, are simulated in this study. In scenario a, 400 agents belong to one single group. In scenario b, the 400 agents are evenly distributed into 20 different groups with 20 agents in each group. Each simulation was run for 200 iterations. The final agent distribution maps are presented in Fig 1(b) and (c). As shown in Fig 1(b), for scenario a, all agents belong to the same group. These agents can freely exchange information about their searching performance. Every agent wants to adopt the problem solution that can generate the highest fitness value. This will cause all agents to swarm around the highest fitness value peak in the landscape. However, because of the dynamic adaptation character of the landscape, the fitness value of the problem solutions around the highest peak will gradually reduce when the number of agents around it increases. For scenario b, as shown in Fig 1(c), limited and noised communication between agent groups causes some agents not to receive the newest information about the best solution that other agents have found. Consequently, agents are distributed relatively evenly around different solution fitness peaks.

The searching performance of these two group scenarios is shown in Fig 2(a), illustrating the average fitness value vs. simulation time step. Initially, scenario a has a higher fitness value than the scenario b, because in scenario a, with the help of social learning, all agents can quickly aggregate around the highest peak in the land-

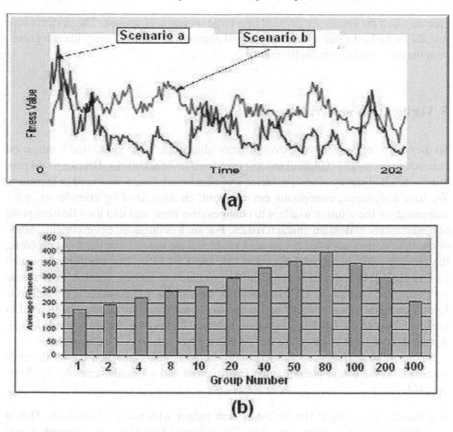

Fig. 2 The comparison of the average fitness values of (a) each simulation iteration for group scenario *a* and *b* (b) whole simulation for different agent group scenarios

scape. However, the fitness value in the landscape will adaptively change according to Eq.7. The congregation of the agents around the highest fitness value will cause a quick decrease of the fitness value of the nearby landscape and eventually cause the sum of the fitness value to quickly reduce. As shown in Fig. 2(a), the profit of scenario a reduces quickly from the peak and remains low. For scenario b, because of the delay and inaccuracy of the information between groups, the agents are evenly distributed around all fitness peaks. This distribution makes the fitness value of the nearby landscape not decrease as quickly as scenario a and maintains a higher group fitness value than scenario a in nearly the whole simulation.To discover the social network architecture that can generate the highest performance, we tested the performance of different group structures that varied from fully connected social network, in which all individuals belong to a single group, to no connection social network, in which no individuals belong to same group. The searching performance is recorded as the average fitness value over the whole simulation. The performance chart is shown in Fig 2(b). According to Fig.2(b), the performance gradually increases when the agents are divided into large numbers of groups. The performance reveals the highest value when there are 80 agent groups and 5 agents in each group, then the performance gradually reduces.

5 Verification and Validation

Studies of the effectiveness of evolutionary algorithms (EA) show that verification requires rigorous and standardized test problems or benchmarks. Branke [4, 6] provides a detailed survey of test problems and benchmark functions found in literature. We have a dynamic, continuous environment, characterized by complexity, a requirement for the solution to adapt to changes over time, and that uses floating point representations of domain characteristics. For such continuous environments, some of the most commonly used benchmark problems are the moving parabola problem, the moving peaks benchmark function [5], and DF1 [12]. Test environments such as DEFEAT [11] can also be used in the verification process for EA in dynamic environments. Our use of DF1 is described in Section 3.1. Another approach uses formal methods to specify and verify swarm-based missions and has been used in verifying the NASA ANTS (Autonomous Nano Technology Swarm) mission [14]. Applicability of this approach to our problem domain is to be determined.

There are several ways to validate agent-based systems and their choice depends on access to the actual phenomenon investigated and on model complexity. They are [16]:

- Compare agent-based simulation/system output with real phenomenon. This is a straightforward comparison, with the difficulty being access to complete real data on the relevant aspects of the phenomenon under study.
- Compare agent-based simulation/system results with mathematical model results. This approach has the disadvantage of requiring construction of the mathematical models which may be difficult to formulate for a complex system.
- Docking with other simulations of the same phenomenon. Docking is the process of aligning two dissimilar models to address the same question or problem, to investigate their similarities and their differences, and to gain new understanding of the issue being investigated [3].

The PASS model research provides a platform for understanding and insights in human self-organized human social groups. With respect to comparison with real phenomenon, it is common knowledge that the collection of real self-organized human community data is difficult. The available data is also questionable because of the potential bias existing when they are collected. Table 1 below lists preferred and optional validation data sources and types. The preferred validation data is real-time data gathered in the domain of interest. If one does not have access to real-time data, then options include use of historical data in the domain of interest, historical data in a parallel domain, or, least favorable, generated data. By parallel domain, we mean one in which the characteristics of the substitute domain parallel, or resemble, those of the domain and phenomenon under study.

Table 1 Preferred and Optional Validation Data Sources and Types

Preferred	Option
Gathered	Generated
Domain of Interest	Parallel Domain
Real-time	Historical

6 Discussion and Conclusion

Most reported searching behavior models only discuss the scenarios in a static environment or a randomly changing environment. The performance evaluation of various approaches is mainly based on how fast an approach can find the optimal point in the benchmark problems. However, the real social world is rarely static and its changes are not random. Most of time, the changes are influenced by the collective actions of the social groups in the world. At the same time, these influenced changes will impact the social groups' actions and structure. In this paper, a modified particle swarm social learning model is developed to simulate the complex interactions and the collective searching of the self-organized groups in an adaptive environment. We constructed a novel agent based simulation to examine the social learning and collective searching behavior of different social group scenarios. Results from the simulation have shown that effective communication is not a necessary requirement for self organized groups to attain higher profit in an adaptive environment.Part of the hesitance to accept multi-agent and swarm-based modeling and simulation results rests in their perceived lack of robustness. The next step in this research will be focused on the model verification and validation.

Acknowledgements Prepared by Oak Ridge National Laboratory, P.O. Box 2008, Oak Ridge, Tennessee 37831-6285, managed by UT-Battelle, LLC, for the U.S. Department of Energy under contract DE-AC05-00OR22725; and by Lockheed Martin, partially funded by internal Lockheed research funds.

References

1. Angeline P. J. (1997): Tracking extrema in dynamic environments. In Angeline, Reynolds, McDonnell and Eberhart (Eds.), Proc. of the 6th Int. Conf. on Evolutionary Programming, LNCS, Vol. 1213 , Springer, 335–345

2. Anthony B., Arlindo S., Tiago S.(2004), MichaelO. N., Robin M. , and Ernesto C.: A Particle Swarm Model of Organizational Adaptation. In Genetic and Evolutionary Computation (GECCO), Seattle, WA, USA 12–23
3. Burton R., (1998): Simulating Organizations: Computational Models of Institutions and Groups, chapter Aligning Simulation Models: A Case Study and Results. AAAI/MIT Press, Cambridge, Massachusetts.
4. Branke, J., (1999): "Evolutionary Algorithms for Dynamic Optimization Problems - A Survey", Technical Report 387, Institute AIFB, University of Karlsruhe .
5. Branke, I., (1999): "Memory Enhanced Evolutionary Algorithms for Changing Optimization Problems", Proceedings of Congress on Evolutionary Computation CEC-99, pp. 1875-1882, IEEE.
6. Branke, J., (2002): Evolutionary Optimization in Dynamic Environments, Kluwer Academic.
7. Cecilia D. C., Riccardo P., and Paolo D. C., (2006): Modelling Group-Foraging Behaviour with Particle Swarms. Lecture Notes in Computer Science, vol. 4193/2006, 661–670
8. Clerc M. and Kennedy J., (2002): The particle swarm-explosion, stability, and convergence in a multidimensional complex space. IEEE Transactions on Evolutionary Computation, vol. 6 58–73
9. Cui X., Hardin C. T., Ragade R. K., Potok T. E., and Elmaghraby A. S., (2005): Tracking non-stationary optimal solution by particle swarm optimizer. in Proceedings of Software Engineering, Artificial Intelligence, Networking and Parallel/ Distributed Computing, Towson, MD, USA 133–138
10. Eberhart R. and Kennedy J., (1995): A new optimizer using particle swarm theory. In Proceedings of the Sixth International Symposium on Micro Machine and Human Science, Nagoya, Japan 39–43
11. Etaner-Uyar, Sima A., and Turgut U. H., (2004): "An Event-Driven Test Framework for Evolutionary Algorithms in Dynamic Environments," IEEE, pp. 2265-2272 .
12. Morrison R. W. and DeJong K. A., (1999): A test problem generator for non-stationary environments. In Proceedings of the 1999 Congress on Evolutionary Computation, Washington, DC, USA 2047-2053
13. Parsopoulos K. E. and Vrahatis M. N., (2002): Recent approaches to global optimization problems through particle swarm optimization. Natural Computing 1 235–306
14. Rouff C. A., Truszkowski W. F., Hinchey M. G., Rash J. L., (2004): "Verification of emergent behaviors in swarm based systems", Proc. 11th IEEE International Conference on Engineering Computer-Based Systems (ECBS), Workshop on Engineering Autonomic Systems (EASe), pp. 443-448. IEEE Computer Society Press, Los Alamitos, CA, Brno, Czech Republic .
15. Tisue S., (2004): NetLogo: A Simple Environment for Modeling Complexity. In International Conference on Complex Systems, Boston, MA
16. Xu J., Gao Y., and Madey G., (2003): "A Docking Experiment: Swarm and Repast for Social Network Modeling," .

Social Network Analysis: Tasks and Tools

Steven Loscalzo[†] and Lei Yu[*]

[†]sloscal1@binghamton.edu, Department of Computer Science, Binghamton University
[*]lyu@cs.binghamton.edu, Department of Computer Science, Binghamton University

Abstract Social network analysis can provide great insights into systems composed of interacting objects, and have been successfully applied to various domains. With many different ways to analyze social networks, no single tool currently supports all analysis tasks, but some incorporate more functionality than others. Moreover, the emergence of a new class of social network analysis techniques, link mining, presents a new range of analysis support to provide by the tools. This paper introduces representative social network analysis tasks from traditional, link mining, and visualization aspects, and evaluates a set of tools with diverse general characteristics and social network analysis functionality.

1 Introduction

Social networks model systems where there is some measurable interaction between multiple objects. The goal of behavior modeling, as stated by Sloane, is to understand, explain, describe, and predict behaviors [10]. As long as the behavior involves some measurable interaction between objects, a social network can be used to describe the behavioral model. Since social networks are naturally encountered in behavior modeling, techniques for social network analysis (SNA) can provide great insights about the behavioral process at the heart of a social network.

To understand social network analysis it is first necessary to learn what makes up a social network. In its most general form, a social network is a model of resource flow between objects [11]. This model typically takes the form of a graph, with vertices corresponding to objects and edges representing resource flow. Vertices are usually referred to as the actors of a social network and edges are known as relations. Social networks may contain more than one type of actors as well as more than one type of relations between actors; these types of networks are called heterogeneous social networks. Additionally, relations may be directed, meaning the relation only makes sense from a source actor to a destination actor, or undirected which shows that the relation is mutual between two actors. Even though the word social implies that actors are people, social networks may be comprised of any entity that takes part in a resource flow.

Social network analysis has been studied in many situations from various domains. A common example of SNA from the marketing domain is the problem of selecting the opinion leader from a group of people to target for a viral advertising campaign [7]. The actors here are people and the relations are the types of interactions between people; for instance, face-to-face, telephone, or online. In the epidemiological domain, SNA can be used to track and predict disease transmission to give clues as to which populations will be at increased risk of infection. As in the previous case, the actors in this social network are people, and the relations are the direct contact interactions between people. Energy companies can use SNA to discover what areas of their grid need to be modified to meet demand, or which stations are most relied upon and so most in need of adequate redundancy systems. The actors here are the different power stations located on the grid, and the relations are the power lines that link them. More examples of SNA applications can be found in [4, 5].

Since SNA has been applied for around 40 years [9], it is in no way surprising to find that there are many tools available that employ some level of SNA functionality. The International Network for Social Network Analysis maintains a long listing of these tools on their website [8]. The focus of this survey is not to rank every available tool, but instead present a set of tools with diverse general characteristics (as discussed in Section 3) and judge their ability to handle representative tasks from traditional, link mining, and visualization aspects of SNA (as discussed in Section 2).

In the rest of the paper, we categorize various SNA tasks and provide a brief introduction to representative tasks in each category in Section 2, and then present our tool evaluation criteria and evaluate selected SNA tools in Section 3. Section 4 provides conclusions based on tool evaluation results.

2 Social Network Analysis Tasks

Social networks can be analyzed in various ways depending on a user's needs. We categorize existing SNA tasks into three broad categories: traditional analysis, link mining, and visual analysis. Traditional analysis tasks are those studied at the dawn of SNA. These tasks focus on analyzing the structural properties of a network by measuring values of actors and relations, such as degree, in-degree, and out-degree measurements for many calculations. Degree is the overall count of links associated with a particular actor in an undirected network, in-degree is a count of links that are incident on a given actor, and out-degree is the number of links that originate from a given actor. In-degree and out-degree are defined only for directed networks. In the following explanations of common traditional SNA tasks, in-degree and out-degree can be replaced with the overall degree of the actor if the network is undirected.

Degree Centrality - The in-degree of a node, any incident link is counted towards this centrality. This is one of the measures that help determine the importance of an actor to a network.

Between-ness - The number of shortest path connections between any two actors in the network that the actor in question lies along.

Closeness - The average shortest path length from an actor to all actors reachable from it. This gives an idea of how close an actor is to other actors in the network. A low closeness measure means that an actor is towards the center of the network, while a high closeness measure means that an actor is only a fringe member.

Eigenvector Centrality - A relative value based on the degree centrality of those actors that are connected to a given actor. An actor has a high eigenvector centrality if many of its out-degree links are to actors with high centrality measures. Intuitively this is like saying that an actor is more important if he/she has important contacts.

K-Core Identification - Actors who have at least k relations among other actors who have at least k relations. This produces groups of actors who are reachable via many paths.

K-Clique Identification - A group of actors where every actor in the group is at most k relations away from every other actor in the same group. This can be used to find decoupled groups as no actor in a group needs to be directly connected to any other actor in the same group, merely k relations away.

K-Clan Identification - A group of actors where any two actors are separated by at most k relations that take place between other actors in the group. This results in closer groups than those identified by k-cliques.

These are a subset of all traditional SNA tasks, and included here because of their popularity and usefulness in link mining tasks. Details about these properties and their calculation as well as other traditional SNA tasks are described in [6, 11].

Link mining tasks have evolved much more recently than traditional SNA tasks due to their reliance on recent advances in data mining techniques. Link mining can be regarded as data mining on social networks, where the traditional classification and clustering tasks of data mining are reformulated to work with network data. Like traditional data mining, link mining tasks can be either predictive or descriptive in nature. A predictive task aims to predict the values of a particular attribute based on the values of other attributes, while a descriptive task aims to derive patterns that summarize the underlying relationships in data. Among the following representative link mining tasks, the first six are predictive tasks, while the last two are descriptive tasks. Detailed explanation of these tasks are provided in [4, 5], and some real-world applications are introduced in [5].

Link-Based Object Classification - Traditional data mining classifies objects based on their attributes. Since objects are now part of a network, link-based classification uses not only attribute information, but also additional link information about existing objects to classify new objects.

Link Existence Prediction - Given a social network at one time, the goal of this task is to predict what link(s) will form when the network is studied at a later time. Take for instance the bibliographic domain, where a social network can be defined as the collection of all authors who have published in a computer science area, and all relations which connect two authors who have collaborated on a paper. Given this network at one time, deciding what pairs of authors will collaborate (gain a link to each other) in the next year is a link existence prediction task.

Link Cardinality Estimation - Similar, but more relaxed than link existence prediction, link cardinality estimation addresses the task of predicting the count of links that a node will have before that node is actually added to the social network, but not which exact links will be added.

Link Type Prediction - This task predicts the type or purpose of a link based on properties of actors. For example, in a social network where actors are either professors or students and relations are defined as either friend, collaborator, or advisor, if a link is formed between two students the task of predicting the type of this link is an application of link type prediction.

Graph Classification - This can be considered as a scaled up version of the typical data mining classification task. In data mining a single object is classified to a particular class, whereas in graph classification the whole network of actors and relations is classified to a single class, as if the collection is a massive object.

Object Reconciliation - As a special case of object classification, object reconciliation attempts to discover whether any actors in a social network actually correspond to the same entity. Citing another example from the bibliographic domain, deciding if two authors who share the same name are in fact the same author is an object reconciliation task.

Group Detection - Similar to traditional clustering tasks, group detection focuses on determining natural clusters of similar actors in a given network. Unlike traditional clustering which decides similarity among objects only based on attributes, the criteria for similarity in group detection take into account both the attributes of actors and the relations among actors.

Sub-Graph Discovery - Deciding the presence of a well-known sub-graph in a given social network is a sub-graph discovery task. In the business domain, assume that research shows that a group of employees are likely engaging in a fraud scheme if they form a specific social network. Knowing this social network, a large company can search through its entire social network to discover sub-graphs of this type which would identify potential fraud activity within the company.

Besides traditional analysis and link mining tasks, visual analysis tasks which revolve around visualizing the information in social networks are common in SNA. While conceptually straightforward, visual analysis enables SNA experts to intuitively draw conclusions about social networks that might otherwise remain hidden even after applying other analysis techniques. Therefore, visual analysis is an important part of SNA. Numerous ways of visualizing the information in a social network are illustrated in [3].

3 Tools Evaluation

The criteria we used to evaluate SNA tools fall into two distinct categories. The first deals with the general characteristics of the tool itself, and the second evaluates each tool's ability to address the SNA tasks introduced previously.

The general characteristics evaluated here focus on tool availability to researchers and practitioners working on social network analysis. Each tool is investigated to determine its latest version, license type, developer type, user interface type, availability of a full version, and availability of source code. The values for each characteristic are as follows.

Version - The listed software version at the time of evaluation.
License Type - The licenses used by these tools mainly consist of Berkeley Software Distribution (BSD), GNU General Public License (GPL), and (F)reeware license. Additionally, Ucinet [2] is under what its user manual [1] describes as a (book) license, meaning that any amount of installations can be made as long as only one instance of the program can be used at a time. DyNet SE and NetMiner (vary) depending on the configuration of the purchased product.
Developer Type - A (C)ompany, a (T)eam of developers, or an (I)ndividual.
User Interface Type - (G)raphical, (B)rowser-based, or code (L)ibraries without predefined user interface. If a tool is a library then the language of the library is noted.
Demo/Full Version - Whether a (D)emo or a (F)ull version of the tool is freely available.
Open Source - Whether the source code of the tool is freely available.

We investigated a large number of SNA tools that were available during this survey, and selected eight tools which exemplify a wide range of diverse general characteristics specified above and support various SNA tasks introduced in Section 2. The names and sources of these tools are listed below, and their general characteristics are summarized in Table 1.

Agna -   http://www.geocities.com/imbenta/agna/
DyNet SE - http://www.atalab.com/newsite/software/dynet/dynet_se.php

JUNG - http://jung.sourceforge.net/
NetMiner 3 - http://www.netminer.com/NetMiner/home_01.jsp
NetVis - http://www.netvis.org/index.php
SNA Package of Carter's Archive - http://erzuli.ss.uci.edu/R.stuff/
StOCHNET - http://stat.gamma.rug.nl/stocnet/
Ucinet - http://www.analytictech.com/ucinet/ucinet.htm

Table 1. General characteristics of selected SNA tools.

	Agna	DyNet SE	JUNG	NM 3[1]	Net Vis	SNA[2]	St[3]	Ucinet
Version	2.11	1.1	2	3.2.0	2	1.2	1.8	6.177
License	F	vary	BSD	vary	GPL	GPL	N/A	book
Developer	I	C	T	C	I	T	T	C
User Interface	G	G	L-Java	G	B	L-R	G	G
Demo/Full Version	F	D	F	D	F	F	F	D
Open Source	N	N	Y	N	Y	Y	Y	N

[1] NetMiner 3

[2] SNA Package of Carter's Archive

[3] StOCHNET

The SNA tasks investigated in our study include all of the traditional analysis, link mining, and visualization tasks described in Section 2. Table 2 summarizes the results of the investigation for each tool in these areas. The possible values are (Y)es and (N)o, denoting whether a tool clearly supports a function or not. Note that group detection is the only link mining task listed in Table 2 since none of the tools included here shows clear support of other link mining tasks.

As shown in Table 2, few of the investigated tools were designed with full generality in mind. On the contrary, most of the tools were created to address a specific task in SNA and then grew to their present state. We now discuss the relative strengths of each of the selected tools.

Agna is designed with casual or non-expert users in mind. Its lightweight GUI makes it easy to use while still supporting some basic but useful SNA tasks.

The most similar tools in this survey are DyNet SE, NetMiner 3, and Ucinet. Each of them was developed by a company with advanced users in mind, and provides a rich GUI environment. In addition to the functions listed in Table 2, all three support a host of other SNA functions and would be well suited to many SNA applications. Ucinet is so far the most mature software product of the three, but its visuals are relatively outdated as compared to the other two.

Table 2. Tasks supported by selected SNA tools.

	Agna	DyNet SE	JUNG	NM 3[1]	NetVis	SNA[2]	St[3]	Ucinet
Degree Centrality	Y	Y	Y	Y	Y	Y	Y	Y
E-vector Centrality	N	Y	Y	Y	N	N	N	Y
Between-Ness	N	Y	Y	Y	N	N	N	Y
Closeness	N	Y	N	Y	Y	Y	N	Y
k-Clans	N	N	N	Y	N	N	N	Y
k-Cores	N	Y	N	Y	N	N	N	Y
k-Cliques	N	Y	N	Y	N	N	N	Y
Group Detection	N	N	Y	Y	N	N	N	Y
Visualization	Y	Y	Y	Y	Y	Y	N	Y

[1] NetMiner 3
[2] SNA Package of Carter's Archive
[3] StOCHNET

JUNG is a Java library and thus has the ability to be easily extended. The main focus of the JUNG project revolves around displaying graphs according to many different layouts, but it does have quite a few algorithms that can be applied and extended for SNA.

NetVis is currently implemented as a web interface for SNA, making it unique among the tools tested. The tool does not need to be installed on any computer; data can be added directly to the web interface and analysis can be conducted accordingly. This feature makes NetVis a valid tool for casual users of SNA who do not have large datasets to work with.

The SNA Package is a library like JUNG except that it is more tailored to traditional SNA tasks and other statistical analysis. Visualization is not directly supported, but being a library it can be worked into a system that does visualization. A unique feature is that it is written in the R language which is not popular with computer science professionals but is commonly used by statisticians.

StOCHNET only provides one of the functions listed in Table 2, but it does have the ability to conduct advanced SNA tasks. StOCHNET focuses on the generation and simulation of social networks and provides functions to evaluate networks on those levels. If this is the desired kind of analysis then StOCHNET is a good choice because it wraps five other tools and so provides a generous range of analysis capabilities.

4 Conclusions

According to the evaluation results in Section 3, both NetMiner 3 and Ucinet performed very well on the evaluated tasks, and either would be well suited to many SNA activities. One limitation of these tools is the fact that they are closed-source tools, which might be too restrictive for some applications. If extensibility is an important consideration for a project, JUNG supports a majority of the representative SNA tasks and might be a good choice.

Perhaps even more important than what the tools are currently capable of is what they are not capable of. Many of the most powerful tasks in SNA, the link mining tasks, are not well supported by the current generation of SNA tools. One of the reasons lies in that effective algorithms for solving many of these tasks in the general case are yet to be developed since link mining is still on the frontier of research. Promising future work would be to find good solutions to these tasks and subsequently implement them in widely available tools.

5 References

1. Borgatti, S P, Everett, M G, and Freeman, L C (2007) Ucinet 5 for Windows: Software for SNA User's Guide. Harvard: Analytic Technologies.
2. Borgatti, S P, Everett, M G, and Freeman, L C (2002) Ucinet 6 for Windows: Software for Social Network Analysis. Harvard: Analytic Technologies.
3. Freeman, LC (2000) Visualizing Social Networks. Journal of Social Structure: vol. 1, number 1.
4. Getoor, L and Diehl, C (2005) Link Mining: A Survey. SIGKDD Explorations: vol. 7, issue 2.
5. Han, J. and Kamber, M (2006) Data Mining: Concepts and Techniques, 2nd ed. Morgan Kaufmann, New York.
6. Jamali, M and Abolhassani, H (2006) Different Aspects of Social Network Analysis IEEE/WIC/ACM International Conference on Web Intelligence: 66-72.
7. Kiss, C, Scholz, A, and Bichler, M (2006) Evaluating Centrality Measures in Large Call Graphs. In: proceedings of the 8th IEEE International Conference on E-Commerce Technology.
8. Richards, W (2007) Computer Programs for Social Network Analysis. http://www.insna.org/INSNA/soft_inf.html.
9. Scott, J (2000) Social Network Analysis: A Handbook 2nd ed. Sage Publications, Thunder Oaks, California.
10. Sloane, H (1992) What is Behavior Analysis. http://www.behavior.org/behavior/what_is_behavior_analysis.cfm.

11. Wasserman, S and Faust, K (1994) Social Network Analysis: Methods and Applications. Cambridge University Press.

Behavioral Entropy of a Cellular Phone User

Santi Phithakkitnukoon[1], Husain Husna[2], and Ram Dantu[3]

[1]santi@unt.edu, Department of Comp. Sci. & Eng., University of North Texas
[2]hjh0036@unt.edu, Department of Comp. Sci. & Eng., University of North Texas
[3]rdantu@unt.edu, Department of Comp. Sci. & Eng., University of North Texas

Abstract The increase of advanced service offered by cellular networks draws lots of interest from researchers to study the networks and phone user behavior. With the evolution of Voice over IP, cellular phone usage is expected to increase exponentially. In this paper, we analyze the behavior of cellular phone users and identify behavior signatures based on their calling patterns. We quantify and infer the relationship of a person's randomness levels using information entropy based on the location of the user, time of the call, inter-connected time, and duration of the call. We use real-life call logs of 94 mobile phone users collected at MIT by the Reality Mining Project group for a period of nine months. We are able to capture the user's calling behavior on various parameters and interesting relationship between randomness levels in individual's life and calling pattern using correlation coefficients and factor analysis. This study extends our understanding of cellular phone user behavior and characterizes cellular phone users in forms of randomness level.

1 Introduction

Mobile phone has moved beyond being a mere technological object and has become an integral part of many people's social lives. This has had profound implications on both how people as individuals perceive communication as well as in the patterns of communication of humans as a society. In this paper we try to capture the behavior of phone users based on their calling patterns and infer trend of behavior dependencies using techniques such as Entropy, principal factor analysis and correlation function. We present a new method for precise measurement of randomness of phone user based on their calling patterns such as location of the call, talk time, calling time and interconnected time and infer relationship among them.

Recently there has been increasingly growing interests in the field of mobile social networks analysis, but due to the unavailability of data, there have been far fewer studies. The Reality Mining Project at Massachusetts Institute of Technology (MIT) [1] has made publicly available large datasets from their projects. We implement our techniques on the Reality Mining dataset which was collected over nine months by monitoring the cell phone usage of 94 participants. The information collected in the call logs includes user IDs (unique number representing a mobile phone user), time of call, call direction (incoming and outgoing), incoming call description (missed,

160

accepted), talk time, and tower IDs (location of phone users). These 94 phone users are students, professors, and staffs.

Using purely objective data first time the researchers can get an accurate glimpse into human behaviors. Our interest in this data set is to study the behavior of the phone user using information theory, data mining, and data reduction techniques.

In [2], the authors attempted to quantify the amount of predictable structure in an individual's life using entropic metric and discovered that people who live high-entropy lives tend to be more random or less predictable than people who live low-entropy lives. This raises the question about how this entropy-based randomness level is related to the randomness level in calling behavior. Does it mean that people who have high-entropy lives also have high-entropy calling patterns? To answer this question, we find it interesting to study the relationship between the randomness level in individuals life and calling pattern.

The main contribution of this paper is to infer the relationship between the randomness levels in behavior of the phone users in a cellular network. We believe that this work can also be extended to predict what services that are suitable for the user.

The rest of this paper is structured as follows: Section 2 carries out the randomness level computation based on entropy. Section 3 discusses the randomness computation result and the relationship among them. The paper is concluded in section 4 with a summary and an outlook on future work.

2 Randomness Level Computation

While individual phone user's calling behavior is random, some users might be more predictable than others. Being more predictable can also mean being less random. To quantify the randomness or amount of predictable structure in an individual calling pattern, the information entropy can be used.

The information entropy or Shannon's entropy is a measure of uncertainty of a random variable. The information entropy as given in Eq. (1) was introduced by Shannon [3], where X is a discrete random variable, $x \in X$, and the probability mass function $p(x) = Pr\{X = x\}$.

$$H(X) = -\sum_x p(x) \log_2 p(x). \tag{1}$$

The calling pattern can be observed from the calling time, inter-connected time (elapsed time between two adjacent call activities), and talk time (duration of call). Let C, I, and T be random variables representing calling time, inter-connected time, and talk time respectively. The entropy of calling time can be calculated by Eq. (2).

$$H(C) = -\sum_{c=1}^{24} p(c) \log_2 p(c), \tag{2}$$

Table 1 Result of Correlation Coefficient

	$H(L)$	$H(C)$	$H(I)$	$H(T)$
$H(L)$	1.0000	0.4651	-0.4695	-0.4642
$H(C)$	0.4651	1.0000	-0.2218	-0.3502
$H(I)$	-0.4695	-0.2218	1.0000	0.2197
$H(T)$	-0.4642	-0.3502	0.2197	1.0000

where the probability $p(c)$ is a ratio of the number of calls during c^{th} hour slot to the total number of calls of all time slots (N).

Similarly, the entropy of inter-connected time can be calculated by Eq. (3) where $p(i)$ is a ratio of the number of inter-connected time whose value is in the interval $[i - 1, i)$ to $N - 1$.

$$H(I) = -\sum_i p(i) \log_2 p(i). \tag{3}$$

Likewise, the entropy of the talk time is given by Eq. (4) where $p(t)$ is a ratio of the talk time whose value is in the interval $[t - 1, t)$ to N.

$$H(T) = -\sum_t p(t) \log_2 p(t). \tag{4}$$

By the same token, the randomness in the individual life's schedule (location), $H(L)$ can also be quantified using information entropy which is defined in Eq. (1).

3 Result and Analysis

Based on our real-life call logs of 94 users, we infer the relationship between the randomness based on the underlying parameters by computing the correlation coefficient [4]. Correlation coefficient is a number between -1 and 1 which measures the degree to which two random variables are linearly related. A correlation coefficient of 1 implies that there is perfect linear relationship between the two random variables. A correlation coefficient of -1 implies that there is inversely proportional relationship between the two random variables. A correlation coefficient of zero implies that there is no linear relationship between the variables. As a preliminary result shown in Table 1, it can be observed that the randomness based on location($H(L)$) and calling time($H(C)$) show high correlation as well as the $H(I)$ and $H(T)$ pair.

Next, we perform factor analysis in order to further study the relationship of the randomness levels (entropy) based on the underlying parameters. The main application of factor analysis is: (1) to reduce the number of variables and (2) to detect structure in the relationship between variables, that is to classify variables [5]. In our analysis we use it for both the purposes. The flow diagram of the principal factor analysis is shown in Fig. 1.

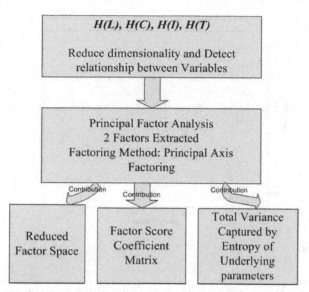

Fig. 1 Flow diagram for principal factor analysis on calculated entropy.

Table 2 Total Variance Explained

Factor	Initial Eigen Values			Extraction		
	Total	Variance(%)	Cumulative(%)	Total	Variance(%)	Cumulative(%)
1	1.59	39.95	39.95	0.92	23.20	23.20
2	1.02	25.61	65.57	0.29	7.26	30.46
3	0.73	18.24	83.81	-	-	-
4	0.64	16.18	100.00	-	-	-

Two principal factors are selected based on the Scree plot [7]. The principal factor plot of the entropy based on four parameters lying on the first and second factor is shown in Fig. 1. It can be observed that the $H(L)$ and $H(C)$ are positively lying on the first factor whereas the $H(I)$ and $H(T)$ are positively lying on the second factor. Since the first and second factor are orthogonal i.e., uncorrelated, one can notice two established relations; one is between $H(L)$ and $H(C)$, and the other one is between $H(I)$ and $H(T)$.

Factor analysis generally is used to encompass both principal components and principal factor analysis. The Eigen values for a given factor measures the variance in all the variables which is accounted for by that factor as stated in Table 2. If a factor has a low eigen value, then it is contributing little to the explanation of variances in the variables and may be ignored as redundant with more important factors.

Eigen value is not the percent of variance explained but rather a measure of amount of variance in relation to total variance (since variables are standardized to have means of 0 and 1, total variance is equal to the number of variables).

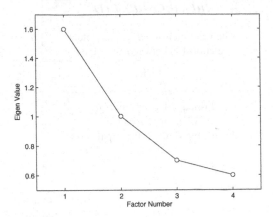

Fig. 2 Scree plot.

Initial eigen values and eigen values after extraction (extracted sums of squared loadings) are same for Principal Component Analysis (PCA) extraction [6], but for factor analysis eigen values after extraction will be lower than their initial counterparts.

Scree plot was developed by Cattell [7] for selecting the number of factors to be retained in order to account for most of the variation. In our analysis, based on Kaiser's criterion [8] the first two factors whose eigen values are greater than 1 (as listed in Table 3) are selected based on the scree plot shown in Fig. 2.

The plot of the entropy based on four parameters lying on the first and second factor is shown in Fig. 3. It can be observed that the entropy based on location and calling time are positively lying on the first factor whereas the entropy based on inter-connected time and talk time are positively lying on the second factor. Since the first and second factor are orthogonal i.e., uncorrelated, one can notice two established relations; (1) between entropy based on location and calling time and (2) entropy based on inter-connected time and talk time.

The scatter plots in Fig. 4 also confirm our findings by showing the proportional relationships between pairs $(H(L), H(C))$ and $(H(I), H(T))$, and inversely proportional relationships among other pairs. The trend (linear-fitting) line is shown in red to emphasize the direction of the relationship, directly proportional (increasing) or inversely proportional (decreasing). Note that the linear fitting is obtained by the least square fitting method [9].

The results based on the correlation coefficients, factor analysis, and scatter plots tell us that there is a high correlation in the randomness in phone user's location and calling time, as well as high correlation in the randomness in phone user's inter-connected time and talk time. This draws the conclusion of our study that phone users who have higher randomness in mobility tend to be more variable in time of making calls but less variable in time spent talking on the phone and the time between connection (idle time). By the same token, the phone users who spend

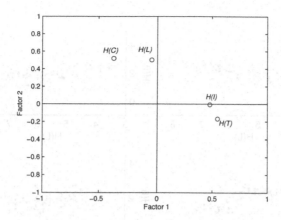

Fig. 3 Principal factor plot.

higher random amount of time talking on the phone (connected time) tend to also be more variable in idle time but not less random in mobility and time of initiating the calls.

We believe that this finding can also be useful for the phone service providers in offering right plans for the right customers based on customer's calling behavior, e.g. suppose that a customer has increasingly high randomness in mobility, service provider might offer this customer a whenever-minute plan which would fit his calling pattern (high $H(L)$ implies high $H(C)$).

4 Conclusion

In this paper, we have presented and analyzed cellular phone user behavior in forms of randomness level using information entropy based on user's location, time of call, inter-connected time, and duration of call. We are able to capture the relationship of the user's randomness level based on the underlying parameters by utilizing the correlation coefficient and factor analysis.

Based on our study, the user's randomness level based on location has high correlation to the randomness level in time of making phone calls and vice versa. Our study also shows that the randomness level based on user's inter-connected time has a high correlation to the randomness level in time spent talking on each phone call.

A knowledge of the randomness levels of a phone user behavior and their relationships extends our understanding in the pattern of user behavior. We believe that this work can also be extended to predict what services that are suitable for the user. This study will also be useful for the future research in this area. As our future direction, this study will be applied to quantify the presence information in terms of willingness level of a phone user in accepting a call.

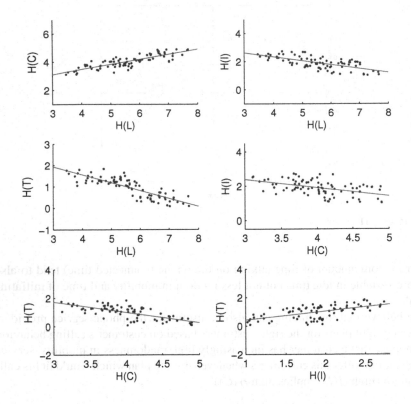

Fig. 4 Scatter plots showing relationships among $H(L), H(C), H(I), and H(T)$ with the linear trend lines.

Acknowledgments.

This work is supported by the National Science Foundation under grants CNS-0627754, CNS-0619871, and CNS-0551694. Any opinions, findings, conclusions or recommendations expressed in this material are those of the author(s) and do not necessarily reflect the views of the National Science Foundation.

We would like to thank the Reality Mining Project group, particularly Dr. Nathan Eagle and Dr. Alex (Sandy) Pentland of MIT Media Lab for providing us the valuable datasets.

References

1. Massachusetts Institute of Technology: Reality Mining Project. Available: http://reality.media.mit.edu/
2. Eagle, N., and Pentlend, A.: Reality Mining: Sensing Complex Social Systems. Personal and Ubiquitous Computing, Vol. 10, No. 4, 2006.
3. Shanon, C.E.: A mathematical theory of communication. Bell System Technical Journal, Vol. 27, pp.379-423 and 623-656, July and October 1948.
4. Cohen, J.: Statistical power analysis for the behavioral sciences. 2nd ed. Lawrence Erlbaum Associates, Hillsdale, NJ, 1988.
5. Jolliffe, I. T.: Principal Component Analysis. 2nd ed. Springer Science+Business Media, 1986, New York USA.
6. Eagle, N., Pentland, A., and Lazer, D.: Infering social network structure using mobile phone data. Proc. of National Academy of Sciences, 2006.
7. Cattell, R. B., and Vogelmann, S.: A Comprehensive trial of the scree and KG criteria for determining the number of factors. Mult. Behav. Res., Vol. 12, pp. 289-325, 1977.
8. Kaiser, H.F.: The application of electronic computers to factor analysis. Educ. Psychol. Meas. 20, 141e151, 1960.
9. De Groen, P.: An introduction to total least squares. in Nieuw Archief voor Wiskunde, Vierde serie, deel 14, 1996, pp. 237-253.

Community Detection in a Large Real-World Social Network

Karsten Steinhaeuser[1] and Nitesh V. Chawla[2]

[1]ksteinha@cse.nd.edu, University of Notre Dame, IN, USA
[2]nchawla@cse.nd.edu, University of Notre Dame, IN, USA

Abstract Identifying meaningful community structure in social networks is a hard problem, and extreme network size or sparseness of the network compound the difficulty of the task. With a proliferation of real-world network datasets there has been an increasing demand for algorithms that work effectively and efficiently. Existing methods are limited by their computational requirements and rely heavily on the network topology, which fails in scale-free networks. Yet, in addition to the network connectivity, many datasets also include attributes of individual nodes, but current methods are unable to incorporate this data. Cognizant of these requirements we propose a simple approach that stirs away from complex algorithms, focusing instead on the edge weights; more specifically, we leverage the node attributes to compute better weights. Our experimental results on a real-world social network show that a simple thresholding method with edge weights based on node attributes is sufficient to identify a very strong community structure.

1 Introduction

Modern data mining is often confronted with the problems arising from complex relationships in data. In social computing, the analysis of social networks has emerged as an area of great interest. On one hand, such interaction networks offer an advantage because they can represent rich, complex information in an intuitive fashion. On the other hand, mining this information can be quite difficult as many existing methods are not directly applicable to network data, and graph theoretic algorithms are computationally very expensive. Therefore, there is an immediate need for efficient algorithms to analyze social networks.

We address one particular task in social network mining, namely *community detection*. A number of methods to address this problem have been proposed, and Newman distinguishes these into two categories: bottom-up "sociological" approaches and top-down "computer science" approaches; a more detailed treatment with examples of each is provided in [5]. Both have been shown to perform well in practice, but regardless of the fundamental approach most algorithms are computationally expensive. Their scalability is limited to at most a few thousand nodes as execution becomes intractable for larger networks [7]. However, datasets containing millions of nodes are becoming readily available, and analyzing them requires highly scalable algorithms.

In this work we take advantage of the important social tendency of *homophily* – "birds of a feather flock together" – to analyze a cellular phone network, which is a unique real-world social network in that it consists of 1.3 million individuals connected by actual communication patterns between them. The search for community structure is guided by a similarity function based on attributes attached to nodes in the network, not just the topology. We believe the latter is limiting as it does not carry the important element of "closeness" among neighbors. Our hypothesis is that using node attributes (in this case demographic information about the individuals) to compute edge weights is sufficient to identify communities in the network, whereas weights computed by other means produce significantly inferior results. We show that a relatively simpler and highly scalable algorithm is able to produce extremely high modularity scores, surpassing empirical limits specified by Newman [6].

The remainder of this paper is organized as follows: In section 2, we describe three different similarity metrics used to weight the edges of a network. In section 3 we present the setup and experimental evaluation on a real-world social network. Finally, in section 4 we conclude with a discussion of the results and their implication for social network analysis.

2 Edge Weighting Methods

We assume that the connectivity of the network is provided as part of the input. If this is all the information given, then the only criterion the algorithm can consider is the network topology, i.e. measurements like clustering coefficients, shortest paths, etc. Yet it is often the case that rich information about nodes and/or edges is available, which allow us to assign them more meaningful weights. In this section, we describe three different methods for weighting the edges of the network: two topological metrics and one based solely on node attributes.

2.1 Terms and Notation

Here we briefly introduce some terms and notation that are used throughout the ensuing discussion. A network is defined as graph $G = (V, E)$ consisting of a set of n nodes V and a set of m edges E between them. Letters i, j, v refer to nodes; $e(i, j)$ denotes an edge connecting nodes i and j, while $w(i, j)$ specifies the weight of the edge. For practical reasons, the graph is represented as an adjacency list such that $neighbors(i)$, the set of all nodes connected to i, is readily accessible. If node attributes for i are available, they are stored as $i.1, i.2, ..., i.r$.

2.2 Clustering Coefficient Similarity (CCS)

Several node similarity metrics are described in [1]. We adopt the topological *clustering coefficient similarity (CCS)* for our work. As the name indicates, the underlying computation requires finding the clustering coefficient CC of node v,

$$CC(v) = \frac{2n_v}{d_v(d_v - 1)}$$

where n_v is the number of triangles v participates in and d_v is the degree of v. The weight $w_{ccs}(i, j)$ is computed as a similarity between two nodes i and j, defined as the difference in their clustering coefficients with (CC) and without (CC') the edge connecting them present,

$$w_{ccs}(i, j) = CC(i) + CC(j) - CC'(i) - CC'(j)$$

Intuitively, CCS measures the contribution of $e(i, j)$ to the connectedness among the immediate neighbors of nodes i and j. Algorithm 1 shows the procedure for weighting the entire graph using this similarity metric.

Algorithm 1 Clustering Coefficient Similarity (CCS)

1: **for** each node $i = 1...n$ **do**
2:     $w(i, j) = 0$
3:     **for** each node $j = 1...neighbors(i)$ **do**
4:         $w(i, j) = w(i, j) + CC(i)$
5:         $w(i, j) = w(i, j) + CC(j)$
6:         remove $e(i, j)$
7:         $w(i, j) = w(i, j) - CC(i)$
8:         $w(i, j) = w(i, j) - CC(j)$
9:         re-insert $e(i, j)$
10:    **end for**
11: **end for**

2.3 Common Neighbor Similarity (CNS)

The second metric is based on a quantity known in set theory as the *Jaccard Coefficient*. For sets P and Q, it is computed as the ratio of the intersection to the union of the two sets. To compute the weight of edge $e(i, j)$, simply substitute $neighbors(i)$ and $neighbors(j)$ for P and Q, respectively, which results in the ratio between the number of neighbors two nodes share (common neighbors) and the total number of nodes they are (collectively) connected to,

$$w_{cns}(i, j) = \frac{|neighbors(i) \cap neighbors(j)|}{|neighbors(i) \cup neighbors(j)|}$$

This metric, called *common neighbor similarity (CNS)*, is intended to capture the overall connectedness among the immediate neighborhood of nodes i and j. Algorithm 2 shows the procedure for weighting the entire graph with the CNS metric.

Algorithm 2 Common Neighbors Similarity (CNS)

1: **for** each node $i = 1...n$ **do**
2:     **for** each node $j = 1...neighbors(i)$ **do**
3:       $w(i,j) = \frac{|neighbors(i) \cap neighbors(j)|}{|neighbors(i) \cup neighbors(j)|}$
4:     **end for**
5: **end for**

2.4 Node Attribute Similarity (NAS)

Note that both of the previous metrics rely solely on the network topology. We postulate that a similarity metric that takes into account node attributes can produce more meaningful weights, thereby improving the community structure. One choice for this scenario might be the Heterogeneous Value Distance Metric [8], but since there is no concept of class among the nodes it cannot be applied directly. However, we can adapt its premise to the situation at hand. We propose to weight edges based on a *node attribute similarity (NAS)* computed as follows: for each nominal attribute a_c, if two connected nodes have the same value then increment the edge weight by one,

$$if\ i.a_c = j.a_c,\ w_{na}(i,j) = w_{na}(i,j) + 1$$

For continuous attributes, to find the weight of edge $e(i,j)$ we first normalize each attribute to $(0,1)$ and then take the arithmetic difference between the pairs of values attribute values to obtain a similarity score. More formally, for each continuous attribute a_n,

$$w_{na}(i,j) + = (1 - \alpha|i.a_n - j.a_n|)$$

where α is a normalizing constant. This metric captures the edge weight as the attribute-similarity of two connected nodes. Algorithm 3 shows the procedure for weighting the entire graph using this heterogeneous NAS metric.

Algorithm 3 Node Attribute Similarity (NAS)

1: **for** each node $i = 1...n$ **do**
2:   **for** each node $j = 1...neighbors(i)$ **do**
3:     $w(i,j) = 0$
4:     **for** each node attribute a **do**
5:       **if** a is nominal **and** $i.a = j.a$ **then**
6:         $w(i,j) = w(i,j) + 1$
7:       **else if** a is continuous **then**
8:         $w(i,j) = w(i,j) + 1 - \alpha|i.a - j.a|$
9:       **end if**
10:     **end for**
11:   **end for**
12: **end for**

Finally, note that it is possible to bypass the edge weighting, for instance if weights are known a priori and provided as part of the input. We give an example of this scenario in the following section.

3 Experimental Evaluation

In this section, we provide an example of a large real-world social network and present an experimental evaluation of the weighting methods discussed above. But first we introduce a validation metric, which allows us to quantify the quality of community structure in networks and enables a comparison between the different weightings.

3.1 Validation Metric

When the true structure of a network is known, the Adjusted Rand Index (ARI) is an appropriate validation metric [4]. In real-world networks this is often not the case, and other criteria must be used instead. Newman and Girvan propose a measure rooted in the notion that nodes within a community should be more tightly connected, while nodes in different communities should share relatively fewer connections [6]. The metric, called *modularity*, takes the fraction of within-community edges minus the expected value of the same number of edges placed at random, summed over all communities. For a network with k communities, it is computed using a $k \times k$ matrix where each element d_{ii} denotes the fraction of edges within community i, and d_{ij} the fraction of edges between communities i and j. Modularity is then given by

$$M = \sum_i (d_{ii} - (\sum_j d_{ij})^2)$$

The value normally ranges from 0 to 1 (higher is better) and can vary widely for different real-world networks. Newman et al. report that in social networks it generally falls between 0.3 and 0.7 [6], but there is no threshold value that necessarily seperates "good" from "bad" community structure.

3.2 Community Detection Method

To identify communities in the network, we first apply one of the edge weighting methods to the network and normalize the edge weights to the range (0,1). We then obtain communities using a simple thresholding method. Given threshold t in the same range (0,1), we place any pair of nodes i and j whose edge weight exceeds the threshold, i.e. $w(i,j) > t$, in the same community.

3.3 Cellular Phone Network

We evaluate our hypothesis that edge weights are a critical foundation to good community structure on a real-world social network constructed from cellular phone records [3]. The data was collected by a major non-American service provider from March 16 to April 15, 2007. Representing each customer by a node and placing an edge between pairs of users who interacted via a phone call or a text message, we obtain a graph of 1,341,960 nodes and 1,216,128 edges. Unlike other examples of large social networks, which are often extracted from online networking sites, this network is a better representation of a true social network as the interaction between two individuals entails a stronger notion of *intent* to communicate. Given its large size, the cellular phone network is quite unique in this regard.

As shown in Figure 1, the degree distribution in the network approximately follows a straight line when plotted on a log-log scale, which is indicative of a power law. This is one of the defining characteristics of a *scale-free network* [2]. Community detection in this class of networks is particularly difficult as nodes tend to be strongly connected to one of a few central hub nodes, but very sparsely connected among one another otherwise. Using topological metrics, this generally results either in a large number of small components or a small number of giant components, but no true community structure. We show that weighting based on node attributes can help overcome this challenge.

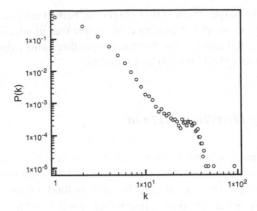

Fig. 1 Degree distribution for the phone network. The presence of a power law indicates that it is a scale-free network.

3.4 Experimental Results

Table 1 shows the effect of threshold t on modularity using the three different edge weighting methods; the execution time for each trial was approximately 40 seconds. We see that this simple thresholding method is sufficient to detect community structure as the modularity values are quite high across the full range of t, although lower thresholds produce better results. In fact, the values over 0.91 for NAS far exceeds the range (0.3,0.7) reported by Newman et al. [6], indicating very strong communities. This shows that the attribute values alone contain some extremely valuable information about the community structure as the NAS metric results in very high modularity.

Table 1 Effects of varying threshold t on modularity with different weighting methods.

Weighting	$t = 0$	0.2	0.4	0.6	0.8
CCS	0.050	0.042	0.020	0.006	0.001
CNS	0.090	0.061	0.022	0.004	0.001
NAS	0.917	0.917	0.917	0.915	0.508

In contrast, the topological weighting methods fail to detect any community structure at all. In this case, the metrics produce many edges with zero weight, fragmenting the network into many thousand singletons and pairs, eliminating the defining characteristics of the community structure.

4 Conclusions

We have explored the viability of various edge weighting methods for the purpose of community detection in very large networks. Specifically, we have shown that edge weights based on the node attribute similarity (i.e. demographic similarity of individuals) are superior to edge weights based on network topology in a large scale-free social network. As witnessed by the fact that a simple thresholding method was sufficient to extract the communities, not only does the NAS metric produce more suitable edge weights, but *all* the information required to detect community structure is contained within those weights. We achieved modularity values exceeding empirical bounds for community structure observed in other (smaller) social networks, confirming that this approach does indeed produce meaningful results. An additional advantage of this method is its simplicity, which makes it scalable to networks of over one million nodes.

References

1. S. Asur, D. Ucar, S. Parthasarathy: An Ensemble Framework for Clustering Protein-Protein Interaction Graphs. In Proceedings of ISMB (2007)
2. A.-L. Barabási and E. Bonabeau: Scale-free networks. Scientific American 288 (2003) 50–59
3. G. Madey, A.-L. Barabási, N. V. Chawla, et al: Enhanced Situational Awareness: Application of DDDAS Concepts to Emergency and Disaster Management. In LNCS 4487 (2007) 1090–1097
4. G. Milligan, M. Cooper: A Study of the Comparability of External Criteria for Hierarchical Cluster Analysis. Multiv. Behav. Res. bf 21 (1986) 441–458
5. M. E. Newman: Detecting community structure in networks. Eur. Phys. J. bf B38 (2004) 321–330
6. M. E. Newman: Finding and evaluating community structure in networks. Phys. Rev. E bf 69 (2004) 023113
7. P. Pons, M. Latapy: Computing communities in large networks using random walks. J. of Graph Alg. and App. bf 10 (2006) 191–218
8. D. R. Wilson, T. R. Martinez: Improved heterogeneous distance functions. J. Art. Int. Res. bf 6 (1997) 1–34

Where are the slums? New approaches to urban regeneration

Beniamino Murgante [†], Giuseppe Las Casas [*], Maria Danese [*]

[†] beniamino.murgante@unibas.it, University of Basilicata, Italy
[*] giuseppe.lascasas@unibas.it, University of Basilicata, Italy
[*] maria.danese@unibas.it, University of Basilicata, Italy

Abstract This paper reports about an application of autocorrelation methods in order to produce more detailed analyses for urban regeneration policies and programs. Generally, a municipality proposes an area as suitable for a urban regeneration program considering the edge of neighbourhoods, but it is possible that only a part of a neighbourhood is interested by social degradation phenomena. Furthermore, it is possible that the more deteriorated area belongs to two different adjacent neighbourhoods. Compared to classical statistical analyses, autocorrelation techniques allow to discover where the concentration of several negative social indicators is located. These methods can determine areas with a high priority of intervention in a more detailed way, thus increasing efficiency and effectiveness of investments in urban regeneration programs. In order to verify the possibility to apply these techniques Bari municipality has been chosen for this research since it shows very different social contexts.

1 Introduction

During the last century, the model of urban development, characterized by urban expansion, has been replaced by the urban regeneration concept. These programs, called urban renewal in some countries, are based on the idea that the space where we will live in the next future is already built [1], therefore it is important to pay attention to heavily populated areas where risks of social conflicts are elevated.

Neighbourhoods built between 1950s and 1960s are characterised by environment deterioration, lack of open space, low availability of car parking, deficiency of street furniture. From a social point of view, great part of immigration is concentrated in these areas, with a high number of unemployed and a low rate of school attendance.

Urban regeneration programs have contributed to encourage the transition from the concept of urban planning focused on town expansion to the qualitative transformation of the existing city in a decisive way. Targets of urban renewal programs can be summarized as quality increase of housing and open space, policy of social cohesion and balance, paying attention to migration phenomena and improvements of area reputation [2].

Various experiences of urban regeneration policy from all over the world differ in contents and forms. Contents can be synthesized as functional integration, social exclusion, environmental degradation, improvement of infrastructure systems. Forms concern community involvement, the possibility to share a program among many towns creating a form of city network in order to reach the same results. More recent forms concern negotiation of new relationships between public administration and private enterprises, using financial resources of private companies, pursuing the need to control public investments. There is a transition from a public administration as a building contractor to a central government stimulating competition among local authorities [3]. The main change is rooted on the shift from an approach based on an huge amount of public funding to a urban renewal through the market mechanism, reinserting areas into real estate market [4]. Methodological and operational innovation introduced in these experiences have encouraged and helped to develop attitude to integration and competition in local authorities. A new assessment system has been introduced in allocating public funds, overcoming the concept of indiscriminate financing and applying reward for quality design, innovation and community needs. The urban regeneration program has introduced also recomposition of socio-economic and financial programming on one hand and land-use planning and urban design on the other.

The designation of Urban Regeneration Program areas generates profound debates. Generally, a municipality proposes an area as suitable for a urban regeneration program, but it considers the edge of neighbourhoods established by bureaucrats. Socioeconomic analysis can account a huge amount of data related to the whole neighbourhood. Nevertheless sometimes it is possible that only a part of a neighbourhood is interested by social degradation phenomena because it could have been designed considering a social mix. In these cases the indicator is diluted and it does not capture the phenomenon throughout its importance. Furthermore, it is possible that the more deteriorated area belongs to two different adjacent neighbourhoods: in this case the municipality will consider an area completely included in a single neighbourhood, even though it may show lesser problems.

Spatial statistics techniques can provide a huge support when choosing areas with high intervention priorities. These methods allow more accurate analysis considering social data at a building scale. In this way areas will be determined considering the spatialization of very detailed data, and neighbourhood limits will be overcome .

This approach has been tested for Bari municipality (Apulia, southern Italy) which is a dynamic trading centre with important industrial activities and high immigration fluxes from Albania and north Africa.

2 An overview of spatial statistics techniques

The main aim of spatial analysis is a better understanding of spatial phenomena aggregations and their spatial relationships. Spatial statistical analyses are techniques using statistical methods in order to determine if data show the same behaviour of the statistical model. Data are treated as random variables. The *events* are spatial occurrences of the considered phenomenon, while *points* are each other arbitrary locations. Each event has a set of attributes describing the nature of the event itself. *Intensity* and *weight* are the most important attributes; the first one is a measure identifying event strength, the second is defined by the analyst who assigns a parameter in order to define if an event is more or less important according to some criteria. Spatial statistics techniques can be grouped in three main categories: *Point Pattern Analysis*, *Spatially Continuous Data Analysis* and *Area Data Analysis* [5].

The first group considers the distribution of point data in the space. They follow three different criteria:

- random distribution: the position of each point is independent on the others points;
- regular distribution: points have a uniform spatial distribution;
- clustered distribution: points are concentrated in clusters.

The second group takes into account spatial location and attributes associated to points, which represent discrete measures of a continuous phenomenon.

The third group analyzes aggregated data on the basis of Waldo Tobler's [6] first law of geography: "Everything is related to everything else, but near things are more related than distant things".

These analyses aim to identify both relationships among variables and *spatial autocorrelation*. If some clusters are found in some regions and a positive spatial autocorrelation is verified during the analysis, it can describe an attraction among points. Negative spatial autocorrelation occurs when deep differences exist in their properties, despite closeness among events [7]. It is impossible to define clusters of the same property in some areas, because a sort of repulsion occurs. Null autocorrelation arises when no effects are surveyed in locations and properties. It can be defined as the case in which events have a random distribution over the study area [8]. Essentially, autocorrelation concept is complementary to independence: events of a distribution can be independent if any kind of spatial relationship exists among them.

Spatial distribution can be affected by two factors:

- first order effect, when it depends on the number of events located in one region;
- second order effect, when it depends on the interaction among events.

2.1 Kernel density Estimation

Kernel density estimation is a point pattern analysis technique, where input data are point themes and outputs are grids. While simple density computes the number of events included in a cell grid considering intensity as an attribute, *kernel density* takes into account a mobile three-dimensional surface which visits each point. The output grid classifies the event L_i according to its distance from the point L, which is the centre of the ellipse generated from the intersection between the surface and the plane containing the events [9]. The *influence function* defines the influence of a point on its neighbourhood. The sum of the *influence functions* of each point can be calculated by means of the *density function*, defined by:

$$\lambda(L) = \sum_{i=1}^{n} \frac{1}{\tau^2} k\left(\frac{L-L_i}{\tau}\right)$$

(1)

where:
- λ is the distribution intensity of points, measured in L;
- L_i is the event i;
- K is the kernel function;
- τ is the bandwidth.

The first factor influencing density values is bandwidth: if τ is too big, then λ value is closer to simple density; if τ is too small, then the surface does not capture the phenomenon. The second factor influencing density values is cell size as in every grid analysis.

2.2 Nearest neighbour Distance

Nearest neighbour distance is a second order property of point pattern analysis and describes the event distribution measuring the distance. This technique analyzes the distance of a point from the nearest source. The events to analyze can be points or cells. The distance between the events is normally defined by the following function:

$$d(L_i, L_j) = \sqrt{(x_i - x_j)^2 + (y_i - y_j)^2}$$

(2)

If $d_{min}(L_i)$ is the nearest neighbour distance for an Li event, it is possible to consider the mean nearest neighbour distance defined by Clark and Evans [10] as:

$$\bar{d}_{min} = \frac{\sum_{i=1}^{n} d_{min}(L_i)}{n}$$

(3)

2.3 Moran index

Moran index [11] allows to transform a simple correlation into a spatial one. This index takes into account the number of events occurring in a certain zone and their intensity. It is a measure of the first order property and can be defined by the following equation:

$$I = \frac{N \sum_i \sum_j w_{ij}(X_i - \bar{X})(X_j - \bar{X})}{(\sum_i \sum_j w_{ij}) \sum_i (X_i - \bar{X})^2}$$

(4)

where:
- N is the number of events;
- X_i and X_j are intensity values in the points i and j (with i≠j), respectively;
- $\bar{X}$ is the average of variables;
- $\sum_i \sum_j w_{ij}(X_i - \bar{X})(X_j - \bar{X})$ is the covariance multiplied by an element of the weight matrix. If X_i and X_j are both either higher or lower than the mean, this term will be positive, if the two terms are in opposite positions compared to the mean the product will be negative;
- w_{ij} is an element of the weight matrix which depends on the contiguity of events. This matrix is strictly connected to the adjacency matrix.

There are two methods to determine wij,: the "Inverse Distance" and the "Fixed Distance Band". In the first method, weights vary according to inverse relation to the distance among events $W_{ij} = d^z_{ij}$ where z is a number smaller then 0.

The second method defines a critical distance beyond which two events will never be adjacent. If the areas to which i and j belong are contiguous, w_{ij} will be equal to 1, otherwise w_{ij} will be equal to 0. *Moran index* I ranges between -1 and 1. If the term is high, autocorrelation is positive, otherwise it is negative.

2.4 G function by Getis and Ord

The G function by Getis and Ord [12] takes into account disaggregated measures of autocorrelation, considering the similitude or the difference of some zones. This index measures the number of events with homogenous features included within a distance d, located for each distribution event. This distance represents the extension within which clusters are produced for particularly high or low intensity values. Getis and Ord's function is represented by the following equation:

$$G_i(d) = \frac{\sum_{i=1}^{n} w_i(d)\, x_i - \bar{x}_i \sum_{i=1}^{n} w_i(d)}{S(i) \sqrt{\left[(N-1)\sum_{i=1}^{n} w_i(d) - \left(\sum_{i=1}^{n} w_i(d)\right)^2 \right] \Big/ N-2}} \tag{5}$$

which is very similar to Moran index, except for $w_{ij}(d)$ which, in this case, represents a weight which varies according to distance.

3 The case of study

In order to verify the possibility to apply these techniques, Bari municipality has been chosen for this research since it shows very different social contexts.

Bari is one of the more developed areas of southern Italy for industrial and tertiary sectors. The location, close to the "heel" of Italy, which historically promoted its trading tradition, today characterizes this area as a sort of gate for migratory fluxes. This municipality has 325.052 inhabitants distributed over 116,20 km².

According to equation (1) all buildings have been represented as points, in order to apply Point Pattern Analysis techniques. The main difference with classical statistical approaches is the possibility to locate each event L_i in the space, by its coordinates (x_i, y_i), in an unambiguous way. An L_i event (equation 6) is a function of its position and attributes characterizing it and quantifying its intensity:

$$L_i = (x_i, y_i, A_1, A_2, \ldots, A_n) \tag{6}$$

The following attributes have been considered in order to calculate kernel density:
- dependency ratio is considered an indicator of economic and social significance. The numerator is composed of people , youth and elderly, who, because of age, cannot be considered economically independent and the de-

nominator by the population older than 15 and younger than 64, who should provide for their livelihood. This index is important in urban regeneration programs because it can determine how many population can provide to building maintenance by themselves;

- foreign population per 100 residents. Normally foreign number is considered as capability attractiveness, but in southern Italy, where concealed labour rate is 22,8% and unemployed rate is 20%, immigration phenomena can be considered a threat and not an opportunity;
- unemployment rate;
- number of people seeking for first job (unemployment rate for young people);
- percent of population which had never been to school or dropped out school without successfully completing primary school programs;
- number of people per room in flats occupied by residents.

All attribute values of various indicators will be standardized according to the following equation: $Z=(X- \mu)/\sigma$, where X represents the value to be normalized, μ is the mean value and σ is the standard deviation. Kernel density has been computed for each attribute and a grid synthetic index can be achieved by summing all kernels (fig.1).

A key factor of kernel density estimation is bandwidth dimension. In order to determine a suitable bandwidth, nearest neighbour distance has been applied. In this case bandwidth dimension is 128 metres while cell size is 10 metres. There are several kinds of kernel functions K (see equation 1); in this case the quartic type has been applied.

Figure 1 highlights two important issues. Highest values (black areas) represent the concentration of several negative social indicators. As supposed in the introduction, this kind of measure has all the limits of the edge of urban regeneration areas based on neighbourhood boundary. The two details on the right part of figure 1 show as areas with high values are located on both parts of neighbourhood boundary and zones which need more urgent interventions are situated between the white line inside the yellow oval.

Moran index is a global measure of spatial autocorrelation which analyzes if and at which degree events are spatially correlated. This index does not give any information about event location. Considering that attributes are connected to buildings, rural areas can produce a sort of mitigation effect. Also, Moran index can be considered as a measure highlighting more concentrated problems determining policy priorities, in the case of a low global autocorrelation degree.

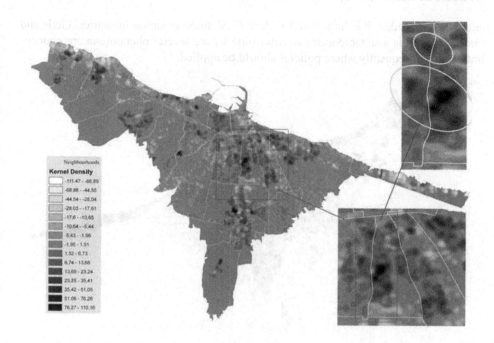

Figure 1. Synthetic index of Kernel Density Estimation of social indicator

Table 1. Moran Index with related distance band for each indicator.

Indicator	Moran's I	Distance band
Dependency ratio	0,170048	40
Unemployment rate	0,285536	30
Number of people seeking for first job	0,227911	40
Foreign population per 100 residents	0,122712	20
Population which had never been to school or dropped out school without successfully completing primary school program	0,71282	20
Number of people per room in flats occupied by residents	0,039907	50

Table 1 shows that education, with very high Moran index, could be a great priority. In urban renewal programs a more diffused school network should be considered. It is impossible to know in which neighbourhood it is better to build a new school only using Moran index. In order to determine the exact location of these problems, further analyses are needed. It is important to adopt local autocorrelation measures. Getis and Ord's function is a suitable index to determine where several phenomena are concentrated and consequently where policies should be applied.

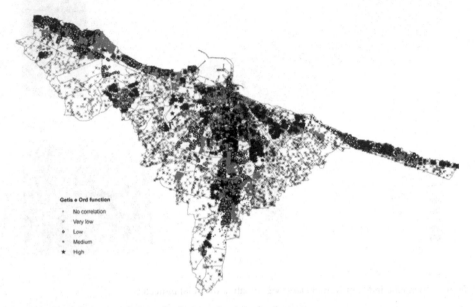

Figure 2. Spatial autocorrelation of primary school graduation rates

Figure 2 shows where high autocorrelation of people with low educational level is located. It is important to pay attention to the class with very low autocorrelation in the external part of the town. A low autocorrelation of people with low educational level can be considered as a medium level of autocorrelation of people with a good educational level. This class is more related with urban sprawl phenomena.

Considering two or more variables at the same time it is possible to achieve other interpretations. Foreigners and unemployed (see figure 3) are not connected with urban sprawl phenomena and these variables are not related. As further remarks, areas with the strongest autocorrelation of immigration are far from zones with high concentration of unemployed. This means that no spatial correlation exists between immigration and unemployment.

4 Final Remarks

One of the frequent accusations which is addressed to policy makers is to fail in understanding the complexity of the city and the real needs of degraded areas [13].

Analyzing urban renewal policies and programs adopted, there is a shift from the first experiences based on emphasizing the built environment to some experiences based on citizen involvement in program choices [14].

This increase in urban renewal quality lacks in defining areas needing more urgent interventions and policies fit to improve neighborhood conditions.

Figure 3. Comparison of medium and high values of spatial autocorrelation of unemployment rate and foreign population per 100 residents

A great support can be represented by the use of spatial autocorrelation methods. These techniques have been adopted in many fields, from epidemics localization [9] to studies on spreading of city services [15], to physical planning processes [16].

These methods can determine areas with high priority of intervention in a more detailed way, increasing efficiency and effectiveness of investments in urban regeneration programs.

The experience explained in this paper is based on simple census data at the urban level and the lack of more specific data limits the number of analyses. In order to develop more interesting considerations, it would be opportune to integrate these indicators during the preliminary phase of urban renewal programs.

References

1. Secchi B. (1984), Le condizioni sono cambiate, Casabella: Architettura come modificazione", n.498/9, Electa Periodici
2. Kleinhans, R.J. (2004), Social implications of housing diversification in urban renewal: A review of recent literature, Journal of Housing and the Built Environment 19: 367–390, 2004. Kluwer Academic Publishers.
3. Murgante B., (2005), Le vicende urbanistiche di Potenza, EditricErmes, Potenza
4. Bonneville M. (2005), The ambiguity of urban renewal in France: Between continuity and rupture, Journal of Housing and the Built Environment (2005) 20: 229–242, Springer
5. Bailey T. C., Gatrell A. C. (1995), Interactive spatial data analysis, Prentice Hall
6. Tobler, W. R., A Computer Model Simulating Urban Growth in the Detroit Region, Economic Geography, 46: 234-240, 1970.
7. Boots B.N., Getis, A. (1988), Point Pattern Analysis, Sage Publications, Newbury Park
8. O'Sullivan D., Unwin D., (2002), Geographic Information Analysis, John Wiley & Sons
9. Gatrell A. C. , Bailey T. C., Diggle P. J., Rowlingson B. S., (1996), Spatial point pattern analysis and its application in geographical epidemiology, Transaction of institute of British Geographer, NS 21 256–274 1996, Royal Geographical Society.
10. Clark P.J., Evans F.C., Distance to nearest neighbour as a measure of spatial relationships in populations, Ecology, 35:445–453, 1954.
11. Moran, P. (1948), The interpretation of statistical maps, Journal of the Royal Statistical Society, n.10
12. Getis, A. and Ord, J. K., (1992), The analysis of spatial association by use of distance statistics, Geo-graphical Analysis, 24, 189-206.
13. Hull A., (2001), "Neighbourhood renewal: A toolkit for regeneration", GeoJournal 51: 301–310, Kluwer Academic Publishers.
14. Carmon N., (1999), Three generations of urban renewal policies: analysis and policy implications, Geoforum 30 145-158, Elsevier Science
15. Borruso G., Schoier G. (2004), Density Analysis on Large Geographical Databases. Search for an Index of Centrality of Services at Urban Scale, in Gavrilova M. L., Gervasi O., Kumar V., Laganà A., Mun Y. and Tan K. J. (eds.), Lecture Note in Computer Scienze Springer-Verlag, Berlin,
16. Murgante B., Las Casas G., Danese M., (2007), The periurban city: geostatistical methods for its definition, in Rumor M., Coors V., Fendel E. M., Zlatanova S. (Eds), Urban and Regional Data Management, Taylor and Francis, London.

A Composable Discrete-Time Cellular Automaton Formalism

Gary R. Mayer[†] and Hessam S. Sarjoughian[*]

[†]Gary.Mayer@asu.edu and [*]Sarjoughian@asu.edu
Arizona Center for Integrative Modeling and Simulation, School of Computing and Informatics, Arizona State University, Tempe, AZ

Abstract Existing Cellular Automata formalisms do not consider heterogenous composition of models. Simulations that are grounded in a suitable modeling formalism offer unique benefits as compared with those that are developed using an ad-hoc combination of modeling concepts and implementation techniques. The emerging and extensive use of CA in simulating complex heterogeneous network systems heightens the importance of formal model specification. An extended discrete-time CA modeling formalism is developed in the context of hybrid modeling with support for external interactions.

1 Introduction

Cellular Automaton modeling has been applied in many ways across many interesting problem domains. A majority of these use the automaton to represent the entire modeled system [1,4]. However, to simulate complex heterogeneous systems, other kinds of models such as rules-based agents are necessary. An example system is one that can model disaster relief efforts. A simulation model for such a system may consist of people, response teams, landscape, and facilities. These can be described using rules-based agent and CA modeling approaches. Synthesis of these different kinds of sub-system models results in a *hybrid* modeled system. A hybrid model is a model of a system, comprised of two or more distinct sub-systems (*e.g.*, [9]). The term "hybrid" is used to imply that each of the sub-system models can be developed with its own appropriate structural and behavioral details. The significance of hybrid modeling is to overcome the unsuitability or impracticality of a monolithic model for describing the disparate types of dynamics (*e.g.*, describing the interactions between fire crews and the spread of fire).

To integrate different kinds of models, a taxonomy of modeling approaches has been proposed. One of these is poly-formalism [6, 10]. In the poly-formalism approach, a separate model is proposed to describe the interactions between the sub-system models. Unlike all other existing model composability approaches, poly-formalism requires a formulation of the models' interactions. This approach specifies the composition of two disparate modeling formalisms via a third modeling formalism. What is gained is a a concise delineation between the sub-system mod-

187

els and their interaction via an *interaction model* [1], which is also called Knowledge Interchange Broker. However, in order to use a CA as a sub-system model within the poly-formalism composed hybrid system model, a CA formalism must distinguish between the interaction between the cells within the CA and between the system as a whole and the interaction model. A CA formalism needs to support the specification of its external input/output with another modeling formalism vice as a black box model. This capability is lacking from current CA specifications.

Model composition aside, it is useful for formal CA models to be implemented in tools such as Geographic Resources Analysis Support System (GRASS) [3]. This requires developing a mapping from CA models to the GRASS implementation constructs. Conversely, a mapping from the GRASS implementation constructs to CA models can enable formalizing the implementation of CA.

To enable the CA to interact with other non-cellular systems in a systematic fashion, the authors have devised a general, domain-neutral description (*i.e.*, formalism) of a multi-dimensional cellular automaton, including its input and its output. The formalism is a network representation built upon the two-dimensional multi-component Discrete-Time System Specification provided in [13]. The base formalism has been expanded to represent the input from and output to other models[2]. It is devised as a 3-dimensional network which may easily be paired down to either a 1-dimensional or 2-dimensional network system.

Considering the system structure from the perspective of a network, it is possible to conceive many different configuration variations — hexagonal, triangular, octahedron, and irregular networks, for example. The formalism defined here could be extended to any regular tessellation of cells, regardless of the "shape" that defines the tessellation. However, for simplicity in describing the formalism, only a grid or cube structure will be explicitly defined. Regardless of network configuration, the connectivity of the network is restricted to immediate neighbors. The cells to which a cell is connected are referred to as its *neighborhood*. A cell's state transition is influenced by its neighborhood and, possibly, by its own current state. The cells which influence the state transition are called the cell's *influencers*. The difference between the set of influencers and the set of cells representing the neighborhood is that the influencers may contain the cell itself.

2 Composable CA Formalism

A *formalism* defines its system using a domain-neutral *specification* and *execution*. The specification is a mathematical description of the system defining structure and behavior. The execution portion defines machinery which can execute model descriptions (*i.e.*, an abstract simulator). A three-dimensional, composable CA, represented by a network, N, can be described by a quintuple:

---

[1] Discussed in more detail in Section 3.

[2] The intent is to compose with non-cellular models but two CA may be mapped as well as described in the example in Section 2.

$$N = \left\langle X_N, Y_N, D, \{M_{ijk}\}, T \right\rangle, \text{ where} \tag{1}$$

$X_N = \{\overline{X}_{ijk} \mid (i,j,k)\} \vee \emptyset$ is the set of input values mapped to the network, N,

$Y_N = \{\overline{Y}_{ijk} \mid (i,j,k)\} \vee \emptyset$ is the set of external output for all M_{ijk} from the network structure, N,

$T = \{h_m \mid 0 \leq m \leq n\}$ is the time-ordered, finite set of time intervals, h_m, (i.e., $\{h_0, h_1, h_2, \ldots, h_n\}$) such that $m, n \in \mathbb{N}_0, \mathbb{N}_0 \equiv (\mathbb{N} \cup \{0\}) - \{\infty\}, h_m \in \mathbb{N}^*$, $\mathbb{N}^* \equiv \mathbb{N} - \{0, \infty\}$ and $\forall h_m, h_m$ occurs before h_{m+1} (i.e., h_0 occurs before h_1, h_1 occurs before h_2, etc.),

$D = \{(i,j,k) \mid a \leq i \leq b, c \leq j \leq d, e \leq k \leq f\}$ is the index set where $a, b, c, d, e,$ $f \in \mathbb{Z}$; the total number of components, $|\{M_{ijk}\}|$, assuming a regular, contiguous network, is $((b-a)+1) \times ((d-c)+1) \times ((f-e)+1)$; and $\forall (i,j,k)$, the component M_{ijk} is specified as

$$M_{ijk} = \left\langle X_{ijk}, Y_{ijk}, Q_{ijk}, \mathrm{I}_{ijk}, \delta_{ijk}, \lambda_{ijk}, T \right\rangle, \text{ where} \tag{2}$$

$X_{ijk} = \dot{X}_{ijk} \cup \overline{X}_{ijk}$ is an arbitrary set of input to M_{ijk}, where $\dot{X}_{ijk}$ is the input into M_{ijk} originating from the set of output of its influencers, I_{ijk}, and $\overline{X}_{ijk} \subseteq X_N$ is the input originating from outside the network, N, which is mapped to M_{ijk}. Note that $\overline{X}_{ijk}$ may be $\emptyset$,

$Y_{ijk} = \dot{Y}_{ijk} \cup \overline{Y}_{ijk}$ is an arbitrary set of output from M_{ijk}, where $\dot{Y}_{ijk}$ is the output from M_{ijk} that acts as input to the cells that M_{ijk} influences and $\overline{Y}_{ijk} \subseteq Y_N$ is the output from M_{ijk} that contributes to the network output, Y_N. Note that $\overline{Y}_{ijk}$ may be $\emptyset$,

Q_{ijk} is the set of states of M_{ijk},

$\mathrm{I}_{ijk} \subseteq D$ is the set of influencers of M_{ijk}; may include M_{ijk}; and $|\mathrm{I}_{ijk}| = 2$ or 3 if both j and k are constant (1-dimensional network), $|\mathrm{I}_{ijk}| = 8$ or 9 if just k is constant (2-dimensional network), or $|\mathrm{I}_{ijk}| = 26$ or 27 if neither i, j, or k are constant (3-dimensional network), the different values for each being indicative of whether or not M_{ijk} acts as its own influencer; furthermore, for any dimension network, $|D| \geq$ the larger of each value for each type of network (i.e., 3, 9, and 27 for a 1-D, 2-D, and 3-D network, respectively),

$\delta_{ijk}(t_r) : \underset{\imath \in \mathrm{I}_{ijk}}{\times} \dot{X}_{ijk}(t_r) \times \overline{X}_{ijk}(t_r) \times Q_{ijk}(t_r) \to Q_{ijk}(t_{r+1})$ is the state transition function of M_{ijk} at time t_r, which is dependant upon the current state,

$Q_{ijk}(t_r)$, to map the set of influencer input, $\underset{\iota \in I_{ijk}}{\times} \dot{X}_{ijk}(t_r)$, and the external

input, $\overline{X}_{ijk}(t_r)$, to the new cell state at time $t_{r+1}, Q_{ijk}(t_{r+1})$,

$\lambda_{ijk} : Q_{ijk} \rightarrow Y_{ijk}$ is the output function of M_{ijk} at time t_r that maps the cell
state, Q_{ijk}, to the component output, Y_{ijk}, and

T is the network time-ordered set of finite time intervals (defined above)
such that $\forall r, 0 \leq r \leq |T| - 1, \{ \delta_{ijk}(t_r) \Rightarrow \Delta t \equiv t_{r+1} - t_r = h_r \in T \}$;
$r, t \in \mathbb{N}_0$; $h_r \in \mathbb{N}^*$; t_r is the time at the current state transition, $\delta_{ijk}(t_r)$; t_{r+1}
is the time at the next state transition; h_r is the time interval between t_r
and t_{r+1}; and the time when the network, N, is in its initial state, q_0, is
represented by t_0.

The component M_{ijk} defined in Eq. (2), which represents a cell within the CA, requires that the output function, λ_{ijk}, occur immediately before the transition function, δ_{ijk}, takes place. Using this approach, if M_{ijk} is specified to be its own influencer, then its output (dependent upon its current state) influences its next state. Otherwise, the next state of M_{ijk} is dependent solely upon the output of the surrounding cells defined to be influencers of M_{ijk}. Also, while the time interval between discrete-time segments is not necessarily constant, the above equation specifies that the difference between any two specific segments, Δt, must be the same for all components, M_{ijk}, within the network at any given time, t_r. In other words, all components within the network execute their output function and undergo state transition at the same scheduled time. This should be apparent from the fact that the network time-ordered, finite set of time intervals, T, is a part of the septuple that defines each component cell within the network.

Note that input and output from the CA is a set of values for each cell within the CA structure. This implies that there must exist a mapping between the input and the CA structure. Similarly, the output has the structure of the network but may then be mapped to whatever form the recipient requires. This mapping is unidirectional and, if between two CA, does not require that the two CA have the same structure. For example, a two-dimensional CA of size 3×3 may be mapped to a three-dimensional CA of size $6 \times 6 \times 3$. The output of the 3×3 network, $Y_{N(3\times3)}$, may be mapped to the input of the $6 \times 6 \times 3$ network, $X_{N(6\times6\times3)}$, by applying the output matrix six times as three quadrants to two of the three z layers of the larger network (see Fig. 1(a)). Note that, in this example, one quadrant of layers 0 and 2 and all of layer 1 of the $6 \times 6 \times 3$ network will not receive external input (i.e., $\overline{X}_{ijk} = \emptyset$). Similarly, the output from the $6 \times 6 \times 3$ network may be mapped to the smaller network as shown in Fig. 1(b). For this example, only the output of the second layer will be mapped to the 3×3 network by partitioning the layer of the $6 \times 6 \times 3$ structure into nine 2×2 sub-networks. For the other layers, $\overline{Y}_{ijk} = \emptyset$. The corresponding output from the cell within each sub-network is then aggregated and provided as input to the respective cell in the smaller network. The input and output in this example is made distinctly different to stress the point that the mapping may be arbitrarily revised as necessitated by the domain. This also highlights a key difference between a multi-dimensional CA and multiple CA which are mapped to one another.

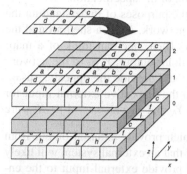
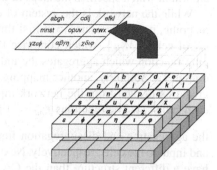

(a) Mapping 2D external output to 3D external input. Cells without letters indicate no external input is received.

(b) Mapping 3D external output to 2D external input. Cells without letters indicate no external output is generated. (Top layer outlined for clarity.)

Fig. 1 An example of mapping two cellular automata

A cellular automaton, regardless of the number dimensions, is a system comprised of homogeneous components. All internal input and output must not be $\emptyset$. On the other hand, two CA mapped to one another may differentiate in many ways — structure, possible states, etc. How they relate to one another is arbitrarily defined via the mapping. Furthermore, the external output from one cellular automaton, Y_N, (and, therefore, the external input, X_N, to the other) may be $\emptyset$.

One other item to consider is edges. In the most simple of cases, the CA may be treated as a torus, in which case, there are no edges. However, while this eases the need to specifically address edges, it does not allow the CA to faithfully represent a realistic landscape. While this paper does not specifically address an execution mechanism for this formalism, it should be said that for edge conditions, it is best that no function explicitly anticipates the number of influencer input values. Instead, it should use the cardinality of the input set. The above formalism does provide the capability to explicitly manage values going into and being output from edge cells. For example, if the CA was being used to model watershed across a landscape, the external input, X_N, could map the specific values entering the edge cells while the external output from each cell, Y_N, could provide the data for material departing the cell (*i.e.*, off of the edge).

Proper execution of the formalism requires more than the state transition and output mechanisms provided in the specification. By specifying that the output function, λ_{ijk}, for each component, M_{ijk}, is solely dependent upon the current state, Q_{ijk}, to produce the output, Y_{ijk}, a delay is introduced which prevents a delayless, infinite feedback loop between components. The use of discrete-time (which does not allow zero-time state transitions) supports the assurance that there will not be a delayless feedback loop once state transitions do occur. Next, a mapping approach for handling the input and output between the cells and the network and between the network and any external system is provided. This section then concludes with an algorithm which specifies the steps in the execution of the specification.

While the network is the system of cells, for the purposes of understanding the mapping, it may be easier to at first think of the network, N, as a shell around the set of cells, $\{M_{ijk}\}$, which comprise the system. Then, one can think of a mapping function which aggregates the individual cellular output, $\overline{Y}_{ijk}$, to the network output, Y_N. Similarly, another mapping function could provide the opposite, disaggregation of data from the network input, X_N, to the respective external input for each cell, $\overline{X}_{ijk}$. The functions $f_{out} : \bigcup_{(i,j,k)\in D} \overline{Y}_{ijk} \to Y_N$ and $f_{in} : X_N \to \bigcup_{(i,j,k)\in D} \overline{X}_{ijk}$ are the aggregation and disaggregation functions which provide external output from and input to the cells, respectively. Next, consider that an external system will likely have a different structure than the CA, may not provide external input to the entire CA structure, may not have external input every execution cycle, and may not receive output from every cell in the network. It would be inefficient and, from a component visibility perspective, inappropriate, for the external system to map to every cell in the network. Thus, two other mapping functions are required. Representing the external system with the symbol, Θ, the output from the network may be given as $g_{out} : Y_N \to \Theta_{input}$ and the input from the external system to the network as $g_{in} : \Theta_{output} \to X_N$. Function composition may then be used to define the entire mapping from the external system, Θ, to the cells and back. In other words, $g_{out} \circ f_{out} : \bigcup_{(i,j,k)\in D} \overline{Y}_{ijk} \to \Theta_{input}$ and $f_{in} \circ g_{in} : \Theta_{output} \to \bigcup_{(i,j,k)\in D} \overline{X}_{ijk}$.

While it may at first appear to have been more efficient to employ two functions which map the external system and cells directly, the reader is reminded that the purpose of this work is to present a "composable CA" which is part of a larger system. To draw away from a monolithic system and to provide flexibility in design, interaction between systems, and reuse, the concept behind the four mapping equations should be employed. These draw a distinct delineation between what occurs within the network and outside of it. Furthermore, it allows the external system to use its own unique identifiers for cells without specific knowledge of the network's internal indexing structure. Referring back to Fig. 1(a) as an example, if the 2D system is considered to be just an arbitrary external system, then the set of output, Θ_{output}, could be considered to be the set of nine values, $\{a, b, \ldots, i\}$, and may be indexed and structured in any manner that is appropriate within Θ. g_{in} would map these nine values to the specific 54 cells in the 3D cellular automaton using the network index set, D. The output, for instance, may be a (CA-index, value) pair. f_{in} would then have the specific knowledge of the network structure to provide the data to the cells specified by the index set. With this approach, the external mapping functions, g_{in} and g_{out}, and, therefore, the external system, only needs knowledge of how the network references its individual cells and not the structure of the network itself. In the case that the poly-formalism approach is used, the g_{in} and g_{out} functions are specified within the interaction model. With this proposed approach, the CA formalism may be realized in the GRASS environment.

A general algorithm for executing the composable CA specification is given as:

1. Retrieve the next time interval, h_m.
2. Execute the output function, λ_{ijk}, of each component, M_{ijk}, in the network.
3. Execute $g_{out} \circ f_{out}$ to generate the external output from the network.
4. Execute $f_{in} \circ g_{in}$ to generate external input to each component, M_{ijk}.
5. Execute the transition function, δ_{ijk}, of each component, M_{ijk}.
6. Increment the current time based upon the time interval, h_m.

3 Realization

As stated in Section 1, a CA has potential benefits as a model for many different systems. The *realization* of each (how the formalism is expressed in terms of software architecture and specific domain constraints) depends on the application domain. While this formalism could be used for stand-alone CA systems, its major benefits are gained in a hybrid system. As such, an agent-landscape environment, devised from the Mediterranean Landscape Dynamics (MEDLAND) project [8], will be used as the exemplar domain. MEDLAND is an on-going international and multi-disciplinary effort. One of the project's goals is to develop a laboratory in which to study humans, the environment, and their relationship. In the simulation laboratory, humans are represented by agents and the environment is represented by a landscape model. The landscape model will be the candidate sub-system to demonstrate a realization of the composable CA formalism. Furthermore, an interaction model will be devised to interact with the CA, providing it with input and capturing output.

The interaction model (IM) composes the agent and the landscape sub-system models. It is within the realization of each sub-system model that the interactions between them occur through data transformation, synchronization, concurrency, and timing. The IM is, in essence, a model of the interaction between the two sub-systems models which has its own formalism and realization separate from the two composed sub-system models. It is through the explicit modeling of the interactions within the IM that interaction visibility and, therefore, management of interaction complexity, is gained [5,6]. The exemplar landscape model will realize this CA formalism in the GRASS environment. As a Geographic Information System (GIS), GRASS is described as a georeferenced software development environment that uses map algebra operations to modify its data. Data is stored in a file called a map in either a raster or vector format. GRASS is comprised of independent modules that can acts upon the maps. The modeler is able to define complex dynamics by executing one or more modules against one or maps either manually or via the use of scripts which contain the predetermined execution sequence. From a formalism perspective, map algebra is too general to provide an effective formal representation of the landscape dynamics that we wish to model. However, the CA specification provided above can be implemented within the GRASS environment by being particular about which GRASS modules are implemented and how they are applied against the mapsets. There is a close resemblance to the regular grid-like tessellation of cells within a GRASS raster map and those within a cellular automaton and it is important to understand how the two data structures relate to one another.

A GRASS raster map may be formatted for either 2-dimensional or 3-dimensional data storage. Each raster map represents one data value (*e.g.*, soil depth). Another value, such as numeric value for land cover, would be stored in a different map. The benefit of using a georeferenced environment like GRASS is that the two maps (assuming they represent the same region) would have the same dimensions, cell resolution, and, therefore, number of cells. Thus, if a cell (x, y) referenced in one would have a representation in the other. Furthermore, GRASS modules are devised such they can perform calculations across maps given these geospatial relationships. In contrast, a cell within a CA has a state. That state can be comprised of many variables (*e.g.*, soil depth **and** land cover). Thus, if a CA is realized within GRASS, the relationship between a cellular automaton cell and GRASS maps is 1-to-many. Also, while a cell may have dimension as part of its state, these values have little impact on the CA as a system. So, to ease the burden of association between the two, it would be beneficial to choose a GRASS region and resolution setting that creates a 1:1 cell count ratio between the GRASS map and the CA index set. Otherwise, a more complex mapping will be required[3].

The exemplar landscape model is a representation of a large watershed area in the early Bronze Age on which farmers tend fields. It is part of a larger system, a hybrid model, which is a composition of a the landscape sub-system model and a farmer sub-system model. The landscape model is itself a composition of three models — land cover, soil erosion, and rainfall[4]. As shown in Fig. 2, these models are interdependent. Soil erosion uses land cover, elevation, and rainfall values to determine how much soil is removed and or deposited in areas. The land cover prevents soil erosion. The heavier the land cover foliage, the lower the erosion in that area. In turn, healthy soil values support larger land cover. If soil is washed away, a smaller amount of foliage is supported. The rainfall model provides the underlying cause of the landscape dynamics. Heavy rains help the land cover grow but also wash away more soil. For simplicity, if the system is modeled over a shorter term, one can assume that the rainfall is constant. Overall, the system operates independently of the farmer model. However, when they are composed using the interaction model, each creates impacts on the other. The farmer, in an effort to create land that is suitable for agriculture, removes some of the land cover. This, in turn, increases the erosion in the area. The reduced soil, combined with the continual farming in the area does not allow the foliage to grow back and soil conditions worsen until farming can no longer be sustained. And, so, the farmer moves to different locations in an effort to

---

[3] The devising of such a mapping is left to the reader's own discretion.

[4] A discussion on the research for the landscape model may be found in [2, 11, 12].

grow enough food to survive. Land cover may grow back but may not be capable of returning to the condition it once was due to changes in soil quality[5].

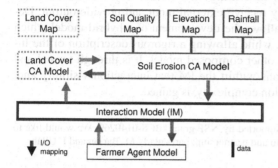

Fig. 2 Exemplar hybrid model.

The dynamics of the CA models can be traced using the algorithm above. For simplicity, we will first assume that each time interval, h_m, has a value of 1. At time t_0, the land cover and soil erosion CA models are initialized. We begin with the first step of the algorithm, retrieving the next time interval, h_0, which is 1. The output function of each CA, λ_{ijk}, then provides output, Y_{ijk}, based upon the initial state of each cell in the respective CA. This occurs both internally as $\dot{Y}_{ijk}$ for every cell that that each influences and externally to create $\overline{Y}_{ijk}$. In step 3 of the algorithm, $\overline{Y}_{ijk}$ is mapped to the input of the receiving system. In the case where the receiving system is another CA, this would be X_N. Step 4 then maps the external input, X_N, to the respective cells, $\overline{X}_{ijk}$, for each CA. This completes the output portion of the execution algorithm. In step 5, each cell in the cellular automata undergoes a state transition based upon the new input values from their influencers, $\dot{X}_{ijk}$, and from external sources, $\overline{X}_{ijk}$[6]. Once the input is accounted for, the cell changes state, as appropriate. The current time of the simulation is then updated by the current time interval, 1, and the algorithm then repeats for time t_1.

The complexity of hybrid model interactions needs to be managed. Even a model as seemingly simple as the one shown in Fig. 2 can be quite complex in its interactions. Each of the land cover, soil erosion, and farmer models may behave in a predictable fashion when tested individually. However, when allowed to interact, the three may exhibit *emergent behaviors* which are not specifically defined in any of the models. To validate the model as a whole, it must be understood why these behaviors are emerging and if they are correct for the inputs being provided across models. This is a difficult task which is made easier through the explicit interaction visibility provided by the interaction model. However, this visibility would be impractical if it were not for a formal composition of the two composed models [6].

---

[5] The specifics of the hybrid model are not the focus of this paper. The reader is encouraged to read [5,7] for more details about the models themselves.

[6] The specifics of things such as priority of internal versus external input and managing receipt of multiple external input is a domain specific and is not discussed here.

4 Conclusion

This paper shows key benefits for describing, in a formal way, the methodology by which the composable CA may receive external input and provide output. Developing a CA to this formalism allows the development of a hybrid model which incorporates a cellular automaton, while allowing a rigorous description of the interaction between the CA and the other composed model. It is through formal and explicit modeling of the interactions within the IM that interaction visibility and, therefore, management of interaction complexity, is gained.

Acknowledgements This research is supported by NSF grant #BCS-0140269. We would like to thank the MEDLAND team for their help and partnership; particularly M. Barton and I. Ullah.

References

1. Anderson JA (2006) Automata Theory with Modern Applications. Cambridge University Press, New York.
2. Arrowsmith R, DiMaggio E, et al (2006) Geomorphic mapping and paleoterrain generation for use in modeling Holocene (8,000 - 1,500 yr) agropastoral landuse and landscape interactions in southeast Spain. In: Am Geophys Union, San Francisco, CA.
3. GRASS (2007) Geographic Resources Analysis Support System. http://grass.itc.it/. Cited Nov 2007
4. Ilachinski A (2001) Cellular Automata: A Discrete Universe. World Scientific, New Jersey.
5. Mayer GR, Sarjoughian HS (2007) Complexities of simulating a hybrid agent-landscape model using multi-formalism composability. In: Proc of the 2007 Spring Simulation Conf (SpringSim '07), 161-168, Norfolk, VA.
6. Mayer GR, Sarjoughian HS (2007) Complexities of modeling a hybrid agent-landscape system using poly-formalism composition. In: J Adapt Complex Syst *(submitted)*.
7. Mayer GR, Sarjoughian HS, et al (2006) Simulation modeling for human community and agricultural landuse. In: Proc of the 2006 Spring Simulation Conf (SpringSim '06), 65-72, Huntsville, AL.
8. MEDLAND (2007) Landuse and Landscape Socioecology in the Mediterranean Basin: A Natural Laboratory for the Study of the Long-Term Interaction of Human and Natural Systems. http://www.asu.edu/clas/shesc/projects/medland/. Cited 01 Dec 2007.
9. Ntaimo L, Khargharia B, et al (2004) Forest fire spread and suppression in DEVS. In: Simulation Trans, 80(10), 479-500.
10. Sarjoughian HS (2006) Model composability. In: Perrone LF et al (eds) Proc of the 2006 Winter Simulation Conf (WinterSim '06), 149-158, Monterey, CA.
11. Soto M, Fall P, et al (2007) Land Cover Change in the Southern Levant: 1973 to 2007. Paper presented at the ASPRS Southwest Tech Conf, Arizona State University, Phoenix, AZ.
12. Ullah I, Barton M (2007) Alternative futures of the past: modeling Neolithic landuse and its consequences in the ancient Mediterranean. In: Proc of 72nd Annu Meet of the Soc for Am Archaeol, Austin, TX.
13. Zeigler BP, Praehofer H, Kim TG (2000) Theory of Modeling and Simulation: Integrated Discrete Event and Continuous Complex Dynamic Systems, 2nd edn. Academic Press, CA.

Designing Group Annotations and Process Visualizations for Role-Based Collaboration

Gregorio Convertino, Anna Wu, Xiaolong (Luke) Zhang, Craig H. Ganoe, Blaine Hoffman and John M. Carroll

{gconvertino, auw133, lzhang, cganoe, bhoffman, jcarroll}@ist.psu.edu
College of Information Sciences and Technology, The Pennsylvania State University, University Park, PA, 16802

Abstract: Team collaboration in situations like emergency management often involves sharing and management of knowledge among distributed domain experts. This study extends our previous research on improving common ground building among collaborators with role-based multiple views, and proposes prototypes of annotation and process visualization tools to further enhance common ground building. We illustrate specific needs through two problem scenarios and propose the designed prototypes through storyboarding.

"The costs of not doing enough in a coordinated way far outweigh the costs of doing it... in a coordinated, better way." *"We are in a position to do so and we should do it".* The EU commission environment spokesperson commenting on the recent forest fire tragedy in Greece, where at least 60 people died. When Greece requested help, nine EU countries responded within 48 hours and all the relief efforts were coordinated through a centre in Belgium (BBC News, Brussels, August 27 2007).

1 Research Focus

The quote above points to the criticality of coordination in emergency management.

Crisis management teams are an example of multi-expert teams making *complex* decisions in *constrained* conditions (e.g., limited time) and under *uncertainty* (i.e., based on incomplete or contradictory evidence). Analogous teams can be found in other work domains [e.g., 18].

Due to the complexity of the tasks, these teams usually consist of members who are specialized experts with *diverse and stable roles*. They contribute different perspectives and different areas of expertise useful to complete the task successfully. Different perspectives imply different goals, background knowledge, and assumptions, while different areas of expertise imply different languages and responsibilities [13]. Specialist knowledge is meaningful and useful in a specific application context. Trained multi-expert teams handle such mapping between knowledge and context implicitly.

The goal for these teams is to make optimal decisions through timely and accurate *sharing and coordination of expert knowledge*. The work is made difficult by the intrinsic complexity of the task (e.g., massive amount of data, dynamic information, limited time, uncertainty), cognitive limitations of experts (e.g., limited capacity and judgment bias), and constraints that the tools (e.g., mixture of mobile and desktop, groupware) and the collaborative setting (e.g., distributed across places and working environments, limited time) impose.

We study ways in which collaborative technology can support decision-making by helping experts to handle the task complexity and by reducing their cognitive limitations or biases. This requires systematic analysis and modeling of behavior.

Over the last three years in our research program on common ground (CG), we have investigated the formation of group structures that regulate knowledge sharing and affect group performance. Specifically, we study the work of small teams of experts performing emergency management planning tasks on maps in distributed and synchronous settings. Building on observations of real practices in the field, we developed a reference task and a laboratory procedure for studying these types of teams [e.g. 6; 8]. We used this experimental model as a test bed for studying regularities in the team process. Our findings are now informing the scenario-based design of tools for knowledge sharing, awareness, and decision-making.

This paper focuses on the design of tools that support knowledge sharing and decision-making in multi-role teams. We present the initial software prototype, which was adopted in our empirical study on emergency management teams [8]. Based on prior research and empirical observations, we isolate key needs of the teams, which we describe through scenarios. On this basis, we develop visualization and annotation tools.

2 Prior Geocollaborative Prototype

Figure 1 displays the user interface of the prototype (also described in [9]). In this paper, we augment this initial prototype with additional tools. The initial prototype features a team map and multiple role-specific maps (Figure 1). Each map displays multiple layers of geographical data. The team map (on the right) is a shared object that is used collaboratively by all the team members. The role-specific maps (on the left), different for each expert, contain unshared role-specific data layers as well as data layers also in the team map (shared data). A toolbar provides navigation and annotation functionality. However, in the experiments we found that the key limitation of this prototype is that, other than publishing information from role-specific map to team map and annotation list, the system does not provide support for information filtering or aggregation and activity monitoring and regulation.

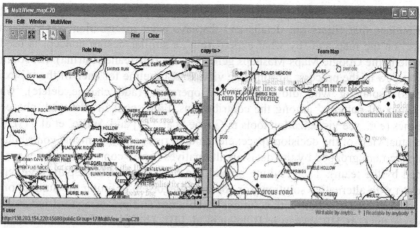

Figure 1: Geocollaborative prototype, showing role (left) and team (right) views along with annotation and telepointer features (also described in [9]).

3 Annotation and Visualization Tools

We propose the introduction of annotation and visualization tools in support of knowledge sharing and decision-making in multi-expert teams.

Prior research on annotations has shown that shared annotations address needs that are related but distinct from the needs addressed by personal annotations [19]. *Group annotation tools* (input or manipulation tools) enable externalizing judgments about information in context. The tagging of the annotations allows post-hoc categorization and bottom-up aggregation of facts or judgments. Annotations also allow for relating different pieces of information or integrating across different media. For example, an annotation in the geographic map can point to another annotation or to role-specific media such as external documents, charts, tables, photos, etc. Process-level annotations help the group manage the decision (e.g., how-to comments, queries on missing information ("?"), proposed categorizations, decisions in progress). These can be re-used in similar situations: e.g., templates that can be reapplied in similar tasks.

Group process visualization tools (output or monitoring tools) enable presenting complex information in ways that can be easily perceived during decision-making (e.g., minimalist map overview, color-coding of roles), give a concrete form to abstract decision-making processes by showing records of actions in relation to geo-space, time, roles, or decision alternative, and reduce the cognitive load (e.g., interactive animations can "link" information pieces spatially dispersed or role-specific). Process visualizations can provide the team with multiple *alternative views* of the decision process. We are exploring three design ideas: (1) A geo-spatial overview of the contributions that filters out details and displays what information was contributed 'where' on the map and 'by what expert' (e.g., a similar design concept is used for example in the Google Maps Mashups generated for Hurricane Katrina). (2) A temporal overview of the process: a n-timeline view (one timeline for each of the 'n' experts or roles) displaying 'who' did what 'type of operation' (add, change, delete), 'when', and 'temporal patterns' among the contributions, time spent in each task phase, and deadlines (e.g., see visualization tools in Ganoe et al. [11] and Andre' et al. [1]); (3) A conceptual view of the decision space: elements of risks and judgments contributed by each expert are grouped by the decision alternatives (or by content categories, or tags, defined by the team). The resulting breakdown of judgments is displayed either in a bar chart with alternative solutions (or content categories) defining the number of bars or a tabular representation with solutions as columns and pro or against judgments as rows (see the CACHE system in Billman et al. [4]). This gives the team a summative balance of the pro or against judgments shared.

Note that it is not our intent to develop new visualization techniques for particular data types. A large amount of research has been done in information visualization to visualize various types of data (e.g., list, 2D data, 3D data, temporal data, etc.). Often visualization designs focus on data type, task type, and individual cognition, but overlook information added during work (e.g., annotations) and group processes, which modulate how the initial data are perceived and how the tasks are performed. We believe that it is not sufficient to see visualization tools just as modules only dependent on data type and task type. It is important to investigate and model the relationship among data types, task type, and group process with and without the mediation of collaborative visualization tools.

4 Identifying Needs based on Previous Research

Evidence from research on group decision-making. Proposition 1: *When using large and dynamic information corpora to make decisions under uncertainty, unsupported individuals and groups are both limited and biased.*
We isolate three main systematic limitations or biases that affect groups' ability to accurately share knowledge and make decisions:

1. Groups' biased analysis and discussion. When sharing and discussing both shared and unshared information, collaborators tend to privilege familiar to unfamiliar information: unless corrective interventions are introduced, group discussions are systematically biased toward shared information to the expenses of unshared and less familiar information (see hidden profile in [17]).

2. The limited capacity in processing massive information in a short interval of time and limited accuracy in distinguishing and managing distinct epistemological categories of information at the same time: facts and sources, inferences about facts and sources, how-to knowledge about the task, experts' related experience (e.g., [12]).
3. Groups' biased synthesis and decisions. When interpreting the meaning and weighting the relevance of information under uncertainty and social influence, group decisions are affected by systematic biases such as anchoring to early hypotheses, confirmation bias, and bias in posterior-probability estimates (e.g., [4]).

Evidence from research on collaborative technology. Proposition 2: *Groups have self-regulation abilities that remain unexploited in face-to-face conditions but that could be exploited through appropriate collaborative technologies.*
Since the early 1980s, various Group Decision-Making Support Systems (GDSS) have been designed with the aim of supporting groups in tasks such as brainstorming, collaborative editing, and meeting recording. Later, the emergence of a new generation of groupware or systems for Computer-Supported Cooperative Work (CSCW) has enabled collaborative tasks in distributed settings and across different devices or platforms. The research on these systems has predominantly focused on how to compensate for the lack of cues characterizing distributed settings.

Only recently has the research started exploring unprecedented opportunities that collaborative technologies offer with respect to face-to-face work conditions. A few recent studies have drawn on models from social psychology to develop tools that can help groups self-regulate their collaborative interactions. DiMicco and collaborators [10] have shown experimentally that shared visualizations that externalize aspects of the communication process can help collocated working groups to communicate more effectively. The presence of the visualization influences the level of participation of the members and the information sharing process during group decision-making. Studies of computer-supported communities have shown that member participation or community development can be influenced using appropriate tools (e.g., [2]).

A few studies on bias in groups have also show that collaborative technology can reduce specific types of bias in decision-making (e.g., [3]). Smallman et al. [16] and Billman et al. [4] have shown that appropriate visualization tools can reduce confirmation bias in intelligence analysis.

Overall, collaborative visualizations and collaborative tools can be used empower groups to process massive amount of data (information coverage and categorization) and provide information about the group process so that the group can become aware of and self-correct biases in the course of the decision-making activity. Situations become more complex in remote cooperation with mobile devices, which is often the case of emergency management work. Biases and limitations can increase because the type of representations and devices are diverse.

5 Addressing Sampling Bias and Limited Capacity

Scenario: *Three experts are planning an emergency management operation: they need to find the best solution available for evacuating a family from a flooded area to a shelter. The three experts, Public Works, Environmental, and Mass Care expert, collaborate through a geo-collaborative system. Each expert has both unique role-specific information and shared information. For example, the Public Works expert is a civil engineer who has critical information on roads, bridges, and public infrastructures. For two of the four available shelters the three experts receive the same information regarding problems of common interest. For the other two shelters they have information regarding problems of role-specific interest. After a full round of sharing about the most prominent issues for each expert, they realize that they have very little time left, thus their group discussion gradually focuses on issues that they have in common. As a result, the information about first two shelters is discussed more exhaustively than the information about the second two shelters. The routes to the second pair of shelters are explored in less detail. Even though they had an equivalent amount of relevant information between the two pairs of shelters, several pieces of critical role-specific information is omitted or ignored in the discussion, which prevents them from finding the best solution. In the last part of the task, after sharing a large number of annotations, the collaborators feel overloaded. The map gets cluttered around the shelters, making it difficult for them to find what they need.*

Storyboarding

Users filter out detailed information added to the map in two possible ways: by level of abstraction and by role. Users can switch among different levels of abstraction from 'Notes', which shows all the text content; 'Dots', which displays only role-based color-coded dots to present where annotations are; and 'Number,' which sums up the total number of annotations by shelter. In the experiment we observed that participants knew their own role-specific information and thus paid selective attention to the information added by their teammates. Users can toggle the display of role information to make the interface clearer (Figure 2, A).

The bar chart of annotation presents an intuitive visualization of the aggregation results. The bar chart (Figure 2, B) shows the information aggregated by shelters. We observed that people compared options one with another in the face to face experiment situation. By showing the annotation categorized by shelters, people may more easily perceive what are the shelters with the least constraints. Moreover, the annotation tool records the annotation and also the users' discussion activity about that. The annotation counter of a specific annotation increases as the users' click on it (Figure 3, A).

By countering and showing the results that some the shelters are over-discussed with respect to others, the system could remind the users of this imbalance.

The timeline graph (Figure 2, C) shows the discussion activity of each expert as time goes by. The squares represent the activity of editing annotations (add and revisit). Different colored squares refer to different shelters. By tracking the process of individual analysis, we can see whether there are 'leaders' who talk a lot in the discussion process and 'followers' who keep silent in the most time, to build a model of the sharing process by the team (e.g., [10]).

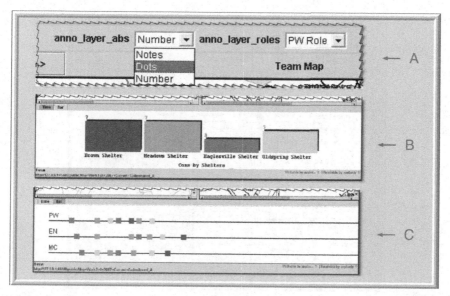

Figure 2. The augmented prototype. A: Filtering tools. B: Bar chart visualization by shelter. C: Timeline visualization by roles.

6 Supporting Aggregation

Scenario: *A group of experts is training to plan potential emergencies on a shared map. The goal is to become increasingly efficient in sharing information and making optimal judgments as a well-coordinated team. When they solve the first emergency scenario the Public Works expert initiates the sharing process by suggesting his preferred shelter and summarizing the infrastructure problems affecting the other three shelters. The Mass Care and Environmental experts follow him by doing the same. All three roles are able to identify the annotations relevant to the shelters by tagging them appropriately. Then the Mass Care expert annotates the map with a "*non-spatial"* *note, where he suggests that they collect in 'summary' annotations non-spatial information on each shelter. This helps the team to quickly compare the four solutions. Thus, as they start the second scenario, the Environmental expert proposes that they start from summarizing non-spatial information shelter by shelter. He suggests: "Let's do the quick summaries as we did the last time", but he would have rather preferred to reuse directly the Mass Care's "*non-spatial" note from the last scenario. Also, he would have preferred to be able to quickly distinguish spatial from non-spatial concerns while they review the information.*

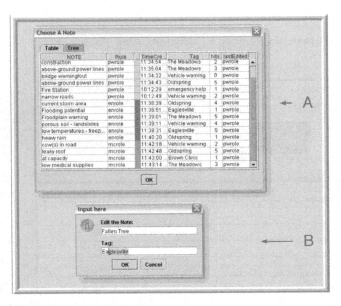

Figure 3. Annotation tools. A: annotation list showing contextual information and tags. B: Tag suggestion via an auto-complete when entering text.

Storyboarding

In addition to adding annotations to the maps, users define tags to classify and categorize these annotations. When editing an annotation, users not only can alter the label but also the tag field of the annotation as in Figure 3, B. By doing so, users can make use of tags, or shared categorizations, as they work together to group information and help coordinate their efforts. As the group works together, the system can give access to previously-used tags to let members leverage prior work, reuse shared conventions, and maintain consistent terminology. Figure 3, B shows a tag suggestion as a user edits the tag field. Observations from the experiment showed that teammates often asked each other which notes were relate to which shelters and, thus, which other notes. Furthermore, groups would add information not related to any specific locations represented on the map that they felt was valuable to the decision-making process.

The annotation tool provides a field for users to enter tags whenever they add an annotation to the map, though a tag is not required for a note to be added to the map. The annotation list tool – which provides information on the creator of the note, the last editor, and the time the note was created – has a field for the notes' tags, allowing for users to edit both the notes' text and their tags (Figure 3, A). As a result, users will be able to group the information that they have provided to one another on the map and better understand the contexts surrounding the notes and how they are inter-related. A benefit of these tags will be the ability for users to group their "non-spatial" information for easier reference regardless of its location on the map.

7 Discussion

An initial prototype for annotation tools has been developed. Currently, users can add annotations, sort them according to role. The visualization tools are in development. We are focusing on aggregating annotations, linking annotations and targets on map, and showing process timelines.

Upon the completion of tools to support group annotations and process visualization, we plan to extend our research into the two directions. First, we will study decision-making in a dynamic information space. Rapidly changing information may increase affect decision-making accuracy and judgment bias. Extending our previous study on complex decision-making in static information environments, we will examine team performance in dynamic environments and the benefits of annotations and process visualization tools through lab experiments. Second, we will develop a theoretical model on team behaviors in collaborative decision-making based on data from experiments. We aim to model team behaviors and common ground (CG) processes in complex decision-making (e.g., key factors and the critical stages of grounding in geo-collaboration), and use this model to further inform the design of collaboration systems. This model can also guide our efforts to design software agents or recommendation systems to assist teams in complex decision-making [15; 10]. Previously, we have developed a preliminary empirical model on common ground processes [5; 8].

Group annotations and process visualization will have a tangible impact on our research on CG. Our interests in knowledge sharing in collaboration go beyond shared mental representations, or shared mental model, per se, and concentrate on the process of constructing such representations, or CG process [14]. Originated from research on language [7]. CG is defined as "the things that we know about what is known by the person we are talking to"[14, p. 270]. We move beyond the scope of communication and model the CG process in the context of computer-supported cooperative work.

Acknowledgements We thank the Office of Naval Research for the support (CKI program, award N000140510549). We thank Helena Mentis, Dejin Zhao, and our colleagues at Penn State.

References

1. André P., Wilson M. L., Russell A., Smith D. A., Owens A. (2007) Continuum: Designing Timelines for Hierarchies, Relationships and Scale, Proc. UIST 2007, Rhode Island.
2. Beenen, G., Ling, K., Wang, X., Chang, K., Frankowski, D., Resnick, P., et al. (2004). Using social psychology to motivate contributions to online communities. In Proceedings of CSCW. New York: ACM Press.
3. Benbasat, I. and Lim, J., 2000. Information technology support for debiasing group judgments: an empirical evaluation. Organ. Behavior & Human Decision Processes, 83, 167–183.
4. Billman, D., Convertino, G., Pirolli, P., Massar, J.P., Shrager, J. (2006). Collaborative intelligence analysis with CACHE: bias reduction and information coverage. PARC, Tech Rep, UIR-2006-09.
5. Carroll J. M., Convertino G., Ganoe C., Rosson M.B. (to appear): Toward a Conceptual Model of Common Ground in Teamwork, In Letsky M. Warner N., Fiore S., & Smith, C., Macrocognition in Teams. Amsterdam, Elsevier.
6. Carroll J.M., Mentis M., Convertino G., Rosson M.B., Ganoe C., Sinha H., Zhao D. (2007) Prototyping Collaborative Geospatial Emergency Planning. Proceedings of ISCRAM 2007.
7. Clark H. H. (1996). Using Language. Cambridge, UK: Cambridge University Press.
8. Convertino G., Mentis H., Rosson M.B., Carroll J.M., Slavkovic, A. (to appear). Articulating Common Ground in Cooperative Work: Content and Process. In Proceedings of CHI 2008 Conference. April 5-10, 2008, Florence, Italy.
9. Convertino G., Zhao D., Ganoe C., Carroll J.M., Rosson M.B.: (2007). A Role-based Multi-View Approach to support GeoCollaboration, Proceedings of HCI International 2007, Beijing, China.
10. DiMicco, J.M., Hollenback, K.J., Pandolfo, A., and Bender, W., (2007). The Impact of Increased Awareness while Face-to-Face. Human-Computer Interaction, 22, 47–96.
11. Ganoe C.H., Somervell J.P., Neale D.C., Isenhour, P.L.,Carroll, J. M., Rosson, M.B., McCrickard, D. S. (2003). Classroom BRIDGE: using collaborative public and desktop timelines to support activity awareness, In Proc. of UIST 2003, 21-30, Vancouver, Canada.
12. Kahneman, D. Slovic P. and Tversky A. (1982). Judgement under uncertainty: Heuristics and biases. New York: Cambridge University Press. 201-208.
13. McCarthy, J.C., Miles, V.C., Monk, (1991). An experimental study of common ground in text-based communication, In Proceedings of CHI 1991, ACM Press, 209-215.
14. Monk, A. (2003). Common Ground in electronically mediated communication: Clark's theory of language use, in J.M. Carroll, Towards a multidisciplinary science of HCI. MIT Press.
15. Nan N., Johnston E. W., Olson J. S., & Bos N. (2005). Beyond Being in the Lab: Using Multi-Agent Modeling to Isolate Competing Hypotheses, Proc. CHI 2005. ACM. 1693-1696.
16. Smallman H. S. (in press): JIGSAW – Joint Intelligence Graphical Situation Awareness Web for Collaborative Intelligence Analysis. In Letsky M. Warner N., Fiore S., & Smith, C., Macrocognition in Teams. Amsterdam, Elsevier.
17. Stasser, G & Titus, W. (2003). Hidden profiles: A brief history.Psychological Inquiry,3-4, 302-11.
18. Stout, R., Cannon-Bowers, J.A., Salas, E. (1996). The role of shared mental models in developing team situation awareness: Implications for training. Training Research J.., 2, 85-116.
19. Weng, C., and Gennari, J.H. (2004). Asynchronous collaborative writing through annotations, Proceedings of CSCW '04, Chicago, ACM Press, 578–581.

Modeling Malaysian Public Opinion by Mining the Malaysian Blogosphere

Brian Ulicny

bulicny@vistology.com, VIStology, Inc. Framingham, MA

Abstract Recent confrontations between Malaysian bloggers and Malaysian authorities have called attention to the role of blogging in Malaysian political life. Tight control over the Malaysian press combined with a national encouragement of Internet activities has resulted in an embrace of blogging as an outlet for political expression. In this paper we examine the promise and difficulties of monitoring a local blogosphere from the outside, and estimate the size of the Malaysian 'sopo' (social/political) blogosphere.

1 Introduction: Blogmining and the IBlogs Project

There are nearly 16 million active blogs on the Internet with more launched every day. Although much – perhaps even the majority -- of what is discussed in the blogosphere is of little consequence and fleeting interest, blogs continue to emerge as powerful organizing mechanisms, giving momentum to ideas that shape public opinion and influence behavior. Recently, Malaysian bloggers have become quite visible in confronting perceived corruption in their national government despite strict governmental control of the major media [3]. Clearly, blogs provide unparalleled access to public opinion about events of the day. Even premier print newspapers like the *New York Times* publish only 15 to 20 of the 1000 letters they receive daily; by contrast, there are over 3000 blog posts that cite a New York Times story every day, including posts not in English [9].

The blogosphere is indeed a great bellwether of changing attitudes and new schools of thought, but only if analysts know which issues to pay attention to and how to identify those issues early in their lifecycle. In beginning to understand the blogosphere associated with a particular part of the world, an analyst may not have much guidance in delimiting a set of blogs whose aggregate behavior matters. Mining a local blogosphere for insight into public opinion requires addressing several difficult issues:

• Which local blogs should be considered?

- How are local blogs to be identified?
- What should blog analytics produce?

Several commercial blogmining services have begun operating in the last few years.[1] These companies offer sophisticated analyses of mentions of (typically, brand-related) entities in blogs, producing analysis of the volume of mentions in social media such as blogs, trends in quantity of mentions, some sentiment analysis (whether a post seemed favorable or unfavorable toward the entity), and some demographic analysis (who is doing the mentioning, by age and other demographic categories). These systems presuppose that there is a set of known entities or issues one wishes to track mentions of, and they do not provide an estimate of the number of bloggers within a demographic category that have not mentioned an entity in a particular category at all. Thus, these services could tell how many bloggers who mentioned X were Malaysian, but not what percentage of the Malaysian blogosphere mentioned X. Further, they cannot list the top entities or issues that are important to the Malaysian blogosphere as a whole.

Similarly, the blog search engines – Technorati, BlogPulse, and others – track the most popular URLs cited by all blogs. BlogPulse can also rank the most cited phrases, blogs, and personal names. These services allow one to search blogs worldwide, and identify the most popular news articles, persons mentioned, or URLs cited worldwide, but they do not allow one to search for blogs within a certain geographic area. Nor do they break down popular mentions or citations by geographic area.

A challenge for understanding the blogosphere, particularly foreign parts of the blogosphere, is determining which blogs warrant attention. In order to measure the temperature of the local blogosphere, as a practical matter, it may be necessary to focus on an appropriate subset of the local blogosphere if trends about significant events are to be identified correctly. Bloggers in any location are too likely to post about ephemeral or merely personal issues to take the entire local blogosphere as the proper object of analysis. But if the entire local blogosphere isn't appropriate, what subset would be more appropriate and more representative? Is the proper object of analysis all bloggers within a certain locale, a subset of bloggers in a certain locale, or just prominent bloggers within a locale, taken individually?

VIStology's IBlogs (International Blogs) project is a three-year effort funded by the Air Force Office of Scientific Research's Distributed Intelligence program to develop a platform for automatically monitoring foreign blogs. This technology pro-

---

[1] These include TMS Media Intelligence/Cymfony (cymfony.com), Relevant Noise (relevantnoise.com), Nielsen BuzzMetrics (buzzmetrics.com), Techdirt (techdirt.com), Umbria (umbrialistens.com), Attentio (attentio.com) and others.

vides blog analysts a tool for monitoring, evaluating, and anticipating the impact of blogs by clustering posts by news event and ranking their significance by relevance, timeliness, specificity and credibility, as measured by novel metrics. This technology will allow analysts to discover, from the bottom up, the issues that are important in a local blogosphere, by providing accurate measurements particular to that locale alone.

Since current blog search engines allow users to discover trends in the blogosphere only by determining the most popular names or news articles or by overall popularity of the blog itself, they tend to favor attention-grabbing stories that may not have lasting significance. The IBlogs search engine, by contrast, ranks blog posts by their relevance to a query, their timeliness, specificity and credibility. Briefly, these are computed as follows (see [8] for details). In particular, because of the exophoric and quotational nature of blogs, it is important to identify links to news articles that posts cite and analyze them. Blog posts are not standalone documents; therefore, information retrieval metrics must take into account the articles they cite as well as the commentary they add.

RELEVANCE: What a blog post is about is determined not only by the text of a post, but also by the text of any news article it references. Terms in news articles and blog posts are not ranked by the familiar tf*idf metric[2] standard in information retrieval in light of the clumpiness of the corpus and journalistic conventions.

TIMELINESS: The timeliness of a blog post is determined by comparing the timestamp of a blog post with the publication date of a news article that it cites. Timeliness, as distinguished from recency, is about proximity to the relevant event.

SPECIFICITY: The number of unique individual entities mentioned in a blog post and any news article it cites determines the specificity of a blog post. This is approximated as the number of unique proper nouns and their variants. Attention is also paid to depth in a domain ontology.

CREDIBILITY: The credibility of a blog author is posts is determined by the presence of various credibility- enhancing features that we have validated as informing human credibility judgments [8, 5, 13]. These include blogging under one's real name, linking to reputable news outlets, attracting non-spam comments, and so on.

This analysis must be computed for each author, since blogs can have multiple authors. The number of inlinks alone does not determine blog credibility.

Architecturally, the IBlogs systems includes a document extraction module, a metrics computing module, an indexer (Lucene), a crawler (Nutch), an ontology reasoner (BaseVISor) and a consistency checker (ConsVISor). See Figure 1.

VIStology's BaseVISor inference engine [11] is used to query domain ontologies. At present, we are using an ontology of Malaysia derived from the 2003 CIA World Factbook. ConsVISor 10 is a rule-based system for checking consistency of ontologies.

---

[2] Term frequency by inverse document frequency.

Figure 1 IBlogs Components

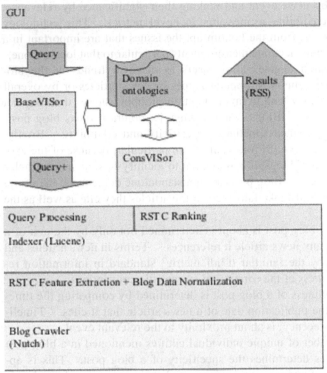

Query Processing	RST C Ranking
Indexer (Lucene)	
RST C Feature Extraction + Blog Data Normalization	
Blog Crawler (Nutch)	

While the idea of blogging involves certain essential features, platforms for blogging are not standardized. That is, blogs do not specify what links constitute their 'blog roll', or which links are 'trackbacks' to other blogs, and so on. Blogs feeds require further analysis because the feed itself is not guaranteed to contain the entire blog post, blog comments, images or profile information relevant to determining blog credibility. All this requires parsing and analyzing HTML blog pages that are designed for human consumption.

IBlogs outputs information annotated according to an ontology of news events and participants. Our goal is to cluster blog posts by the news events that they are about, where any given news event may have more than one news story that reports it, and each of those stories may be published at one or more URLs. A news event is thus typically two levels removed from a blog post that references it.

2 Background: The Malaysian Blogosphere

Malaysia is a nation of approximately 26 million citizens in an ethnically and religiously plural constitutional monarchy in Southeast Asia comprised of thirteen sultanates. Malaysia is a former British colony, and Malaysian English is the second-most popular language, used widely in business. The official religion is Islam, with Muslims constituting roughly 60% of the population. Ethnic Malays form a plurality of the society, constituting 52% of the society, with substantial Chinese- and Indian-descendent populations, as well. Longstanding political tensions exist between these ethnic and religious groups.

The Malaysian blogosphere provides an interesting set of properties for analysis. The Malaysian government tightly controls the press. In 2007, Reporters Without Borders (RSF), the respected journalism watchdog group, ranked Malaysia 124th out of 169 countries surveyed in its 2007 Press Freedom Index [1], its worst rank since the Index was inaugurated in 2002. On the other hand, the Malaysian government has encouraged the use of Web technologies and the growth of Internet-based enterprise. As a result, the Malaysian blogosphere has become important as a place to air dissenting views and publish incriminating information. The Malaysian political blogging community has managed to wage a campaign against perceived governmental corruption.

Malaysian law enshrines free speech as a matter of principle, but the Malaysian sedition law makes restrictions on what can be said about the government or about ethnic groups. The government must license print newspapers in Malaysia, and it has sometimes repealed publishing licenses to newspapers for printing controversial stories. Malaysian defamation law allows prosecution for astronomical damages. Malaysian assembly laws require permits for any gathering in public places (see [15] for details). Organizations, even affiliations of bloggers, must have public, named officers.

Malaysian government officials often issue dismissive comments about bloggers. More seriously, a defamation suit was filed in January 2007, against two of the most prominent Malaysian bloggers [3]. Although the defamation charge has not been specified publicly, it is widely thought that the bloggers are accused of falsely accusing an official newspaper of plagiarizing a well-known American novelist in an editorial. Another blogger was detained under the Official Secrets Act, allegedly for posting a doctored photo of the deputy prime minister. Police complaints have been filed against blogs for allegedly insulting Islam and the King of Malaysia, the elected monarch. Thus, the stakes are high for Malaysian bloggers, and it would be expected that this environment would result in distinct behaviors in the Malaysian blogosphere.

3. Identifying the Malaysian Sopo Blogosphere

In order to take the temperature of the Malaysian blogosphere, it would first be necessary to delineate what constitutes such a collection. It is difficult to identify blogs by location. Blogs are not typically geotagged. Blog directories are very incomplete, and the top commercial blog search engines do not generally enable search by either author or topic location. Geolocating bloggers is not possible by IP address (although commenters may be), because blog platforms allow one to post remotely. In fact, part of the appeal of blogging is the idea of an Internet-based publishing platform that provides as much anonymity as one would like along with the ability to log in to the blogging platform from virtually anywhere. Blogging platforms are used internationally, by Malaysian bloggers as well as bloggers worldwide, particularly Google's Blogger service (www.blogger.com, blogs hosted at blogspot.com), one of the oldest such platforms, and one that is offered without charge. Finally, although blogging syndication is somewhat standardized among a handful of standards (RSS 1.0, RSS 2.0 and Atom), blogging syndication frameworks do not typically embed geotags.

How big is the Malaysian social/political (or 'sopo', as Malaysians term it) blogosphere? Any attempt at predicting the behavior of a delimited Malaysian social/political blogging community (or one associated with some other country or region) will have to be able to delimit some reasonable approximation of the community automatically. One online survey estimates that 46% of those online in Malaysia have a blog [4], which would amount to approximately 6.3 million bloggers given current estimates of internet users[3]. This is based on an online survey at a blog hosting service's website (MSN Live), however, and seems far too large.

Alternatively, the most popular blogging platform in Malaysia allows geographic search of bloggers. A search on Blogger.com returns 152,000 Malaysia blog profiles. Similar queries on the fee-based Typepad blog platform identifies only about 40 additional blogs; Wordpress reveals another 90 bloggers, 920 are listed on The Star's blog directory at AllMalaysia.info[4], 113 are identifiable via MySpace profiles, 185,000 on Facebook, and at least 100,000 on Friendster.[5] Thus, depending on how one delineates blogs, the potential number of bloggers in Malaysia is in the hundreds of thousands.

With some upper bounds on the number of Malaysian bloggers, we then turn to the

---

[3]   Malaysian   Communications   and   Multimedia   Commission,   2006. http://www.internetworldstats.com/asia/my.htm

[4] The Star (Kuala Lumpur) is an official Malaysian newspaper, with a blog directory at AllMalaysia.info/ambp.   Site accessed November, 2007.

[5] All figures as of November, 2007.

question of what constitutes a sopo blog. Can they be detected automatically? Clearly, social/political bloggers, of the kind whose response to events of the day are worth monitoring, are bloggers who monitor the events of the day closely. Therefore, one would expect that they would have a larger than average tendency to cite news sources in their blog than average bloggers.[6]

The data for the Malaysian blogosphere bears this out. We crawled two sets of blogs to identify their likelihood of hyperlinking to online news organizations. The first set of average Malaysian bloggers was identified by taking a random set of bloggers identified as being located in Malaysia from Blogger.com. The second set of bloggers was 77 English-language, crawlable blogs from the Sopo-Sentral blog directory (sopo-sentral.blogspot.com). Sopo-Sentral is "(an attempt to create) a Directory of Blogs & Websites on Society, Politics, and Economy of Malaysia." 559 blogs are listed as primarily social/political (as of April, 2007). The site is associated with the All-Blogs or Malaysian Bloggers Alliance.

Table 1   Citation Properties of Malaysian Bloggers

	SoPo bloggers (77 initial)	Avg. bloggers (50 initial)
Posts and Trackbacks	1754 (219 unique sites)	745 (148 unique sites)
News hrefs	285 (16.2%) (28 unique)	22 (2.9%) (17 uniqe)
Other hrefs	401 (22.0%) (162 unique)	237 (31.8%) (111 unique)

The data in Table 1 shows that the average Malaysian blogger posts somewhat less than the average SoPo blogger. More strikingly, the average SoPo blogger is 5.5 times more likely to link to a news source than the average Malaysian blogger in general. The average blogger is far less concerned with the news of the day, it would seem.[7] Independent news sources dominate in the sopo blogger citations.

A crawl of the Sopo-Sentral directory at a depth of 4 levels produces a set of 220,320 unique URLs representing 4,693 unique sites. Eliminating self-linking

---

[6] For our purposes, we identified a site as a news site if it (1) had an associated print newspaper or magazine; (2) had an associated radio or television station or network; or (3) regularly published selections from newswire services in an appropriately labeled section of its site; and (4) didn't use a commercial blogging platform.

[7] 'A-list' (highly prominent) US political bloggers tend to cite news sources at a much higher frequency (at least 50% of posts, see 2), but this may be a symptom of greater freedom of the press in the U.S.

(links from a site to itself), we used a social network analysis package, SocNetV [6], to analyze the hyperlinking patterns of sites in this network. The network was not very tightly connected. The group indegree centrality metric (see [14]) is 0.0220642, where 0 would indicate that every blog was linked to exactly the same number of blogs, and 1 would indicate that one blog was cited by every other blog. Thus, the Malaysian sopo blogosphere is rather diffuse in its attention. There were 41 classes of actor indegree centrality. The mean indegree actor centrality was 1.72, with an indegree variance of 12.8. This means that a small number of sites had a high degree of inlinking, while most sites had only the minimum number of inlinks. The most inlinked actors were Google advertisement servers.

Table 2:   Bloggers with Highest Actor Indegree Centrality

Blog	AIDC	AIDC stdized	Description
http://edcgoldtruths.wordpress.com	0.0093976	0.54699	Splog
http://rockybru.blogspot.com/	0.0081161	0.47240	Sopo
http://www.jeffooi.com/	0.0059803	0.34808	Sopo
http://kickdefella.wordpress.com/	0.0042716	0.24863	Sopo
http://harismibrahim.wordpress.com	0.0040580	0.23620	Sopo
http://bigdogdotcom.wordpress.com/	0.0040580	0.23620	Sopo
http://csmean.blogspot.com/	0.0038445	0.22376	chinese language blogger
http://generating-revenue-from-your-site.blogspot.com/	0.0032037	0.18647	non-Sopo
http://elearningtyro.wordpress.com/	0.0032037	0.18647	non-Sopo
http://zorro-zorro-unmasked.blogspot.com/	0.0029901	0.17404	Sopo
http://nursamad.blogspot.com/	0.0029901	0.1740	Sopo

As the most central sopo bloggers, we find Rocky (Ahiruddin Attan) and Jeff Ooi, the president and vice president of the Malaysian All-Blogs blogger association, respectively. Ahiruddin Attan writes the blog "Rocky's Bru" (rockybru.blogspot.com) and helped form the blogger's alliance as a solidarity organization in light of the recent blogging push-back. Attan averages 1.9 posts per day. According to Technorati, his blog attracted inlinks from 488 unique sites in last 90 days with 3082 total trackbacks [7].

Jeff Ooi is regarded as the dean of Malaysian bloggers. Ooi's blog Screenshots (www.jeffooi.com) covers a wide variety of issues. Screenshots ranks highly for inlinks with links from 975 unique sites over the past 90 days and 5,465 trackbacks. A major area of concern in recent posts is the Malaysian judiciary and allegations that the judiciary is not as independent as it should be. Ooi averages 3 posts daily.

Other central bloggers are film maker and activist Pak Sheih (kickedefella.wordpress.com), lawyer Haris Ibrahim (harismibrahim.wordpress.com), and Nuraina Samad (nursamad.blogspot.com), a veteran journalist. Two of the remaining top ten sopo bloggers provide no personal details. Only one of the top ten blogs under a pseudonym. Although Malaysian bloggers often post using a nickname or pseudonym, most of the bloggers listed on sopo-sentral provide a real name in their profile.

Of the complete dataset, the top 18 sites are either infrastructure sites like blog advertisement servers or spam blogs. Only 37 sites have an Actor Inlink Degree Centrality (AIDC) of over 20; 78 over 10. 475 (or roughly 10%) have an AIDC over 3; 974 sites have an AIDC over 2. This means that only about 20% of the sites crawled are cited by more than one other blog. So, roughly 500 to 1000 bloggers constitute the active Malaysian sopo blogosphere, with a small, very active core of 75 to 100 bloggers.

These numbers are consistent with data from Blogpulse, which shows that blog posts containing "Malaysia", "Malaysian" or "Malaysians" range between 500 and 1100 posts daily, but issues important to Malaysian sopo bloggers, such as the recent Bersih and Hindraf rallies in Malaysia, generate, respectively 1080 total posts for the Bersih rally, with a daily high of 268 posts, and 1527 total posts for the Hindraf rally, with a daily high of 257 posts. These are roughly consistent with the findings above.

Figure 2 Important Recent Malaysian Blog Topics vs "Malaysia" or "Malaysian(s)"

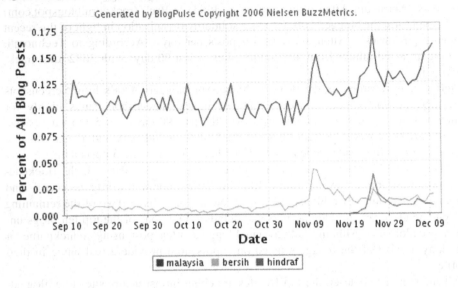

4 Discussion

In approaching a foreign blogosphere, it is difficult to know a priori how many blogs one must analyze in order to accurately measure the temperature of local social/political bloggers. In this study, we have used a directory of social/political bloggers in Malaysia as a point of comparison for presenting evidence that the behavior of social/political bloggers is different from that of ordinary bloggers in their engagement with news sources, even where the press is tightly controlled. Secondly, by using the blog directory as a seed for crawling the local blogosophere, and using social network analysis, we found that the number of active Malaysian social/political bloggers is on the order of 500 to 1000 blogs, rather than the potential millions suggested by another study. This number shows that the temperature of the local social/political blogosphere should be determined by automated means, since 500 to 1000 blogs is too many for a single analyst to follow daily.

Acknowledgments This material is based upon work supported by the United States Air Force under Contract No. FA9550-06-C-0023. This material expresses the views and conclusions of the authors alone and does not necessarily reflect the views of the United States Air Force.

References

1 Reporters Without Borders, 2007 Press Freedom Index. http://www.rsf.org/article.php3?id_article=24025

2 Adamic, L, and N. Glance. 2005. "The Political Blogosphere and the 2004 U.S. Election: Divided They Blog". Proceedings of 3rd International Workshop on Link Discovery. Chicago, IL.

3 Analysis: Tension Between Malaysian Bloggers, Authorities Appears To Intensify FEA20070914318786. OSC Feature - Malaysia -- OSC Analysis 13 Sep 07

4 Colette, Matt. 2006. Blogging Phenomenon Sweeps Asia. MSN Press Release. November 27, 2006.

5 Kaid, L.L, M. Postelnicu. 2007. Credibility of Political Messages on the Internet: A Comparison of Blog Sources. In M. Tremayne (ed.), *Blogging, Citizenship, and the Future of Media*. Routledge: New York.

6 Kalamaras, D. SocNetV Project. Http://Socnetv.sourceforge.net

7 Sifry, D. 2006. Blogging Characteristics by Technorati Authority. http://www.sifry.com/alerts/Slide0006-8.gif

8 Ulicny, B., K. Baclawski, A. Magnus. 2007. New Metrics for Blog Mining. *Proceedings of SPIE Defense & Security Symposium* '07 Vol. #6570. Orlando, FL.

9 Feyer, T., *Editors' Note; The Letters Editor and the Reader: Our Compact, Updated.* New York Times, May 23, 2004

10 Kokar, K, C. Matheus, J. Letkowsk, K. Baclawski and Paul Kogut, *Association in Level 2 Fusion.* In Proc of SPIE Conference on Multisensor, Multisource Information Fusion, Orlando, FL., April 2004 (vistology.com/consvisor)

11 Matheus, C., K. Baclawski, M. Kokar. *BaseVISor: A Triples-Based Inference Engine Outfitted to Process RuleML and R-Entailment Rules.* In Proceedings of the 2nd International Conference on Rules and Rule Languages for the Semantic Web, Athens, GA, Nov. 2006. BaseVISor is freely available (vistology.com/basevisor).

12 Ooi, Jeff. 2007. "Bloggers Sued in Malaysia." Screenshots (blog). http://www.jeffooi.com/2007/01/ bloggers_sued_in_malaysia.php.

13 Ulicny, B., and K. Baclawski, *New Metrics for Newsblog Credibility*, Proceedings of 1st International Conference on Weblogs and Social Media (ICWSM'07). Boulder, CO

14 Wasserman, S, and K. Faust. 1994. Social Network Analysis: Methods and Applications. New York: Cambridge University Press.

15 Wu, Tang Hang. 2006. "Let a Hundred Flowers Bloom: A Malaysian Case Study on Blogging Towards a Democratic Culture." Paper presented at the 3rd Annual Malaysian Research Group In UK and Eire at Manchester Conference Centre, 4 Jun 2005.

Reading Between the Lines: Human-centred Classification of Communication Patterns and Intentions

Daniela Stokar von Neuforn [*] and Katrin Franke [**]

[*] stokarvo@fh-brandenburg.de, Brandenburg University of Applied Sciences, Institute Safety and Security, Germany
[**] kyfranke@ieee.org, Norwegian Information Security Laboratory, Gjøvik University College, Norway

Abstract Author identification will benefit from the deeper understanding of human's approach to infer the intention of an author by some kind of text-content analysis, informal speaking by "read between the lines". This so-called qualitative text analysis aims to derive purpose, context and tone of a communication. By determining characteristics of text that serve as information carrier for communication intents as well as possible interpretations of those information it is assumed to be feasible to develop advanced computer-based methods. So far, however, only few studies on the information carrier and communication intends are performed. In addition, there is no public benchmark dataset available that can support the development and testing of computational methods.

After giving an overview of the current state of art in research and forensic author identification, this paper details a case study on the identification of text-based characteristics and their linkage to specific individuals or target groups by using indicator patterns of human communication and behaviour. Communication patterns determined in this pilot study can be adopted in several other communication contexts and target groups. Moreover, these pattern found will support further research and development of computational methods for qualitative text analysis and the identification of authors.

1 Introduction

The categorization of language by extracting communication patterns is an important means for identifying authors in the prevention, investigation and prosecution of crimes. Criminal cases that call for author identification are widespread as the 2006 published statistic of 4500 cases recorded by the German Federal Police Office (Bundeskriminalamt (BKA)) illustrates (see table 1) [1].

Author identification in the context of criminal investigation uses linguistic methods for classifying communication patterns and individual features. For this purpose "handmade" linguistic analysis was tried to implement into software programmes in order to construct an automatic identification of authors or target groups. These programmes are able to identify text-based characteristics by counting their individual number, forming and positions. Beside this mainly quantitative interpretation of text-based characteristics it is possible to identify the intention of the author by content analysis in order to "read between the lines". This qualitative method determines the intent of a communication.

Table 1: Illustration of the criminal act 2006 in Germany, recognized by the German Federal Police Office, department for identifying authors [1].

Crime Category	Proportion
Extortion	45%
To form a criminal association	15%
Fire raising	12%
Threat	5%
Murder and attempted murder	5%
Industrial espionage	2%
Incitement of the people	2%
Libel	2%
Falsification of documents	2%
Sexual assault, rape	2%
Computer sabotage	2%
Insult	2%

Social behaviour and social context are reproduced in individual communication and are represented in communication patterns. Communication patterns involve the coding and decoding of communication that depends on factors differing within target groups, such as gender, culture, education or the experience of the participants. They determine the perception of certain text-based characteristics, but also their individual interpretation, i.e. which relational relevant indicators are extracted from which text-based characteristics. However, behaviour patterns, which are not directly quantifiable by communication patterns, were not considered in computer-based approaches so far. The following primary factors are comprehensible reasons for using only quantitative linguistic methods in case of forensic detection of authors or target groups: (1) Communication patterns depends on a large number of social (gender, age, status) and contextual (time, surrounding) factors, (2) Production and perception of communication and behavioural pattern differ in target groups and situations, (3) Individual production and perception of communication and behavioural patterns are not usable for forensic detection.

Moreover, the actual practice of computer-based text analysis is the use of software, which is not able to add more text-based characteristics for identification. The program KISTE [1] for example can only identify quantitative information out of text-based communication (number or position of words, sentences, mistakes) [1]. For the purpose of identifying authors, however, it would be more effective to add qualitative information about the author or a target group, like the probability of emotions, interest, status or gender. Both, quantitative and qualitative information connected in a learning algorithm should upgrade the likelihood-ratio of statements about the authorship. Furthermore, it would be possible to prognosticate criminal acts in the nearest future [2] or extract indicators about the emotional and psychological condition of the author.

In a previous study [3] it is assumed that a closer look at the individual language reception of the participants in virtual learning environments can shed light on how participants perceive the text based communication within their virtual learning event. This enables us to draw conclusions for the perception of text-based characteristics and its relationship to the individual factors of text production expressed in the using of text-based characteristics. Hence it is questionable, by means of which text-based characteristics within text-based communication emotional proximity is transported and created. A continuative question contains the individual weighing of the identified characteristics, i.e. their importance for the individual perception. At the same time the individual weighing and interpretation of the identified characteristics allows to draw conclusions about the indicators that the authors use for the assessment of a communication situation.

Our research focuses this specific area of qualitative text analysis. Object of our studies is an extension of the text-style analysis, in such a way that communication patterns are related to the expectations of the listener (information recipient), their social context and group behaviour. In order to support the development of computational methods, a benchmark dataset shall be established by means of methods from social and linguistic sciences. For this purpose, individual characteristics were defined and clustered to derive key features which than give an indication about the emotions and personal characteristics of the sender during writing a text, in our study an email.

The remaining part of this paper is organized as follows: Section 2 gives an overview on current practice in forensic author identification and points to most recent research activities in the field. Section 3 describes the method chosen to establish a

---

[1] In German: Kriminaltechnisches InformationsSystem Texte [1]

benchmark dataset of communication patterns. Results and Discussions are described in Section 4. Finally, Section 5 concludes the discussions and provides pointers for further research and development.

2 Related Work

In the section of pre-crime, patterns of communication are used to recognize behavioural profiles by extracting rules of individual text-based patterns and perpetrator profiles. The most important technology for profiling criminals and terrorists is via data mining and machine-learning algorithms. These are used to cluster data collections and to automate the manual process of searching and discovering key features. Investigative data mining is the visualization, organization, sorting, clustering and segmenting of criminal behaviour, using such data attributes as age, previous arrests, modus operandi, household income, place of birth and many other kinds of data points [2]. Many of this data points are located and expressed in the individual form of communication, which includes patterns of indicators for identification.

Categorization language by extracting communication patterns is necessary for identifying authors in criminal contexts. The identification of authors uses linguistic methods for determining incriminated text in regard to specific individual characteristics. In the past stylo-metric features ("style markers") [4] for the identification of authorship were used in the design of machine-learning algorithm. Ahmed Abbasi and Hsinchun Chen created *Writeprints*, "a principal component analysis based technique that uses a dynamic feature-based sliding window algorithm, making it well suited for visualizing authorship across larger groups of messages from three English and Arabic forums by means Support Vector Machines (SVM)[5]." *Writeprints* is supposed to help identifying authors based on their writing style. Writing style patterns were identified and sorted in four categories that are used extensively for online communication (compare Table 2).

Table 2: Writing-style Features for analysis of online messages as proposed by Zheng et al. [6].

Lexical features	Syntax features	Structural features	Content-specific features
Total number of words	Formation of	Organization of the text	Key words
Words per sentence	sentences	Layout of the text	
Word length distribution	Punctuation	Greetings	
Vocabulary richness	Function/stop	Signatures	
	words	Characters per sentence	
		Characters per word	
		Usage frequency of individual letters	
		Paragraph length	

In the end *Writeprints* is strong for identifying a group of messages, giving additional information into authors writing patterns and authorship tendencies, but it is not powerful enough for authorship identification of individual messages. To provide against terrorism the National Science Foundation starts the Dark Web Project in January 2007, located at the Artificial Intelligence Lab at the University of Arizona. The strength of this project belongs to the integration of Social Network Analysis (SNA) [7].

One of the very few agencies that uses computer assistance for author identification in daily forensic casework is the German Federal Police Office (BKA). The program KISTE [2] contains currently 4500 incriminated texts together with their administrative data (perpetrator, the act, kind of text), linguistic preparation (token, type, kind of words, grammar, syntax), results (mistakes, stile, interpretations, search function), and statistics (type-token-ratio, lexis, length of sentences and text, background statistics). The process of identifying an author with KISTE is divided into three steps: (1) text analysis, (2) text comparison and (3) text collection. Text analysis is the main method for identifying an author. Text comparison can be used if a few reference texts by the suspected author or target group are available. Nevertheless, the focus in the investigation is on text analysis that comprises the analysis of mistakes and the analysis of style. Mistake analysis is further subdivided into identification of mistakes, description of mistakes, genesis of mistakes and interpretation of mistakes. Analysing the text style, so far, means to inspect characteristics that can be quantified as the number of punctuation marks, orthography, morphology, syntax, lexis, structure, and text layout [1].

For the purpose of identifying authors it would be more effective to add qualitative information about the author or a target group, like the probability of emotions, interest, status or gender. However, a more detailed style analysis is challenged by the fact that two different, individual perspectives are involved. The author may vary his/her style in producing a text while the recipient appreciates the style according to his/her own expectations. Therefore, the characteristics of style should be interpreted in an area of conflict between appreciation and different alternatives of communicative behaviour. Furthermore, it shall be emphasised that communication is differently perceived and interpreted according to the context. Factors of authors' behaviour and communication are differing within target groups, such as gender, culture, education or the experience of the participants, determine the perception of certain text-based characteristics, but also their individual interpretation, i.e. which relationally relevant indicators are extracted from which text-based characteristics.

In order to establish a systematic approach for inferring personal motivations and emotions of authors under various conditions a case study in a well observable context/environment seems to be appropriate. Keeping the overall objective of a computer-based approach for qualitative text analysis in mind, a method needs to be chosen that supports the implementation of software routines later on. Consequently, a method that combines qualitative and quantitative measures has to be selected.

3 Method

Two methods were used to conduct the survey. (1) For collecting communication patterns we chose an unaided recall according to Fiske et al. [8, 9], since it provides the possibility to collect a broad overview of text indicators and their intuition perceived by the message recipient. Moreover, the ranking of indictors can be also used for deriving feature categories. (2) For the analysis of the questionnaire we followed the reduction model by Maying [10] that combines a qualitative and a quantitative approach. An overview of the primary working phases of the survey is displayed in Figure 1.

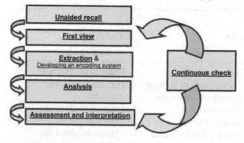

Figure 1: Expiration of the context analysis [12].

More detailed, within the framework of different online university courses it was evaluated, which characteristics of the text-based language are used in the process of individual language reception, to give the particular recipient information about the state of the sender or the character of the communication situation. During numerous presence lectures students were asked in an open questionnaire about the characteristics of text based communication within emails. Without giving examples, these students were asked, by means of which text based characteristics they would draw conclusions about the character and state of their respective communication partner [11]. The characteristics should be spontaneously noted and as far as possible illustrated with examples or evaluations. Using these examples the participants were asked to explain in bullet points how the mentioned characteristics are interpreted (negatively, positively, etc.), i.e. what information is derived from these text based characteristics.

It was then possible to relate the mentioned text based characteristics to different weightings and interpretations of certain verbal expressions using the positioning within the questionnaire, the added examples as well as the frequency of certain quotations. In this process it was assumed that expressions mentioned at the beginning were important for the formation of impression and at the same time were noticed and evaluated first by the participants. The participants also mentioned, which indicators are described and evaluated by the text-based characteristics. These indicators are appreciation, time, social background, relational aspect and degree of trust, status and role, atmosphere, competence and gender are perceived by the respective recipient according to the qualitative evaluation through the existence or the extent of text based characteristics. In summary, the method used for this survey was based on the continuum model of the formation of impression [8, 9]. The goal of this survey was to define text-based characteristics that give information about the relational aspect of messages. 128 female and 128 male German students participated in this survey. The participants were on average 26 years old and were studying in their 5th semester and named around 6 characteristics on average.

Table 3: Illustration of the identified text based characteristics in the four language areas.

A Outer appearance of the message	B Syntax	C Lexik	D Empathical communication
- Address - Length of the message - Closure/greeting - Capital and small writing - Layout - Latency of the answers - Topic - Pictures, colors, graphics - Font - Subject - Attachments - Signature - Answers in the email - Time	- Length of the sentences - Sentence construction - Spelling - Questions - Grammar	- Writing Style - Formal way of writing - Colloquial language - Abbreviations - Adjectives, Adverbs - Slang, Chat language - Extensions - Foreign words - Filling words - Offensive language - Nicknames - Vocabulary, Eloquence - Metaphors - Repetitions - Conjunctive	- Emoticons, symbols - Punctuation marks - Answers - Content (only technical information) - Questions about the well being - Jokes - Highlighting - Compensated phonology - Formal / informal Address - Irony - Personal writing - Reasoning of the message - Apologies

4 Results and Discussion

The notations from the 256 suspects obtained (1535 in total) were distributed over 47 characteristics, which can be separated into 4 clusters or accordant language areas (A-D) of text based communication (Table 3).

In addition to the identification of text-based characteristics, the indicators that are used for the individual perception of the conversation situation could be extracted and associated on the basis of the qualitative evaluation of the additionally given examples and interpretations. That way certain text-based characteristics were either positively or negatively reviewed. Positive reviews were described and associated by the participants with interest, good mood, competence of the sender or with the trust and privacy. On the other hand negative reviews were declared as disinterest, distance, bad mood or incompetence of the sender.

The naming of the indicators by the participants illustrates, what information is obtained from purely written messages, in order to not only evaluate the learning situation, but also mainly evaluate information about the sender of the message according to their individual value system. That way the participants gain information from the written messages about the status, competence or the style. In Figure 2 the identified indicators for the evaluation of the conversation situation and their manifestation in text-based characteristics are illustrated.

In summary, these indicators for social and personal background as well as emotional and contextual information about the situation and characteristics of the author could be added to objective linguistic features. In most cases these added information were not usable for forensic provableness. But the information, based on the specific perception and production of text-based communication in emails could be added to objective linguistic methods into a database in order to arise statistical values about the probability of forensic identification of authors. As another result, emails that were observed continuously could be analyzed in regard of changing emotions or communication features. Forecasting of criminal acts with a specific percentage should be possible.

5 Conclusion and Further Direction

In the presented research it becomes clear that text based language contains a great deal of information about the relationship between text-based communication partners. Beside the written word, there is information hidden "between the lines", which is sent partly consciously (emotions) and partly unconsciously (length of the sentence). Text-based characteristics as the length of a message, the presence of greetings or the answering of questions in short time are mentioned as indicators for the interest and the mood of the communication partner and in the same time a factor of individual perception of the specific communication situation. This research can contribute to the design of communication by explicitly formulating text-based characteristics, which can be considered as essential clues in the interpretation of the interpersonal aspect within written communication.

Despite different scientific procedures the results of his survey are identical to the linguistic techniques of Stylometry, which is the statistical analysis of writing style. Writing-style features are perceived by recipients as indicators of individual characteristics of the author. As a result of this fact, first this study enables us to connect the identified characteristics with emotions and mannerism of individuals. Second, the number of text-based characteristics can be added to the specific perception of indicators of a target group in order to design a benchmark-data set.

In addition to the general illustration of the characteristics used for the evaluation of the context in which the email was written and for the characterization of the author a conclusion about the individual assessment of these characteristics could be drawn. The described approach selected a large kind of text-based characteristics that could give an indication about the identity of the author. That means, language as a medium for the creation of closeness, acceptance, and appreciation or for the signalling of interest shapes the individual communication style and gives an indication of key features, which fit to special target groups or persons.

Further step should be the scientific investigation of produced text in regard to review the already identified indicators and in the same time to scan different target groups in order to get more individual communication patterns for building up a database.

Until now there exists some kind of text mining programmes, which are not very effective for criminal investigation. In an interdisciplinary study the results of content analysis, linguistic and applied computer sciences will be connected with social science in order to evolve to a machine-learning algorithm for classifying patterns of communication and intention of perpetrators. For example, machine-learning algorithms can be used for such questions as what kind of communication patterns gives an indication for which key features in which target group or what are the communicative characteristics of a specific perpetrator. Further machine-learning software can segment a database into statistically significant clusters based on a desired output, such as the identifiable characteristics of suspected criminals or terrorists.

References

1 Sabine Ehrhardt: Sprache und Verbrechen – Forensische Linguistik im Bundeskriminalamt, Ringvorlesung zum Jahr der Geisteswissenschaften, Stuttgart, 21.Mai 2007.
2 Jesús Mena: Investigative Data Mining for Security and Criminal Detection, Butterworth Heinemann / Elsevier Science (USA), 2003, ISBN:0-7506-7613-2.
3 Stokar von Neuforn: Geschlechtsstereotype Rezeption textbasierter Kommunikation in virtuellen Lernumgebungen, Shaker Verlag GmbH, Aachen, ISBN: 978-3-8322-5778-1, 2006.
4 De Vel, O., Anderson, A., Corney, M., Mohay, G.: Mining E-Mail content for author identification forensics. SIMOD Record, 30(4), 2001,55-64.
   www.sigmod.org/record/issues/0112/SPECIAL/6.pdf
5 Ahmed Abbasi, Hsinchun Chen: Visualizing Authorship for Identification, Department of Management Information Systems, The University of Arizona, Tucson, AZ 85721, USA,2006
6 Zheng,R., Quin, Y., Huang, Z., Chen, H.: A Framework for Authorship Analysis of Online Messages: Writing-style Features and Techniques. Journal of the American Society for Information Science and Technology 57(3), 2006,378-393.
7 Dark Web Terrorism Research, http://ai.arizona.edu/research/terror/index.htm, Oct. 2007.
8 Fiske, Susan T., Neuberg, Steven L. (1990). A continuum of impression formation from category-based to individuating processes: Influences of information and motivation on attention and interpretation. In: Advances in experimental social psychology. Hg. Mark P. Zanna, S. 1-74.New York: Academic Press.
9 Fiske, Susan T., Neuberg, Steven L. (1999). The continuum model: Ten years later. In S. Chaiken and Y. Trope (Hrsg.), Dual process theories in social psychology (S.231 – 254). New York: Guilford.
10 Mayring, Philipp : Qualitative Inhaltsanalyse. Grundlagen und Techniken, Weinheim,1988
11 Lamnek, Siegfried (1993). Qualitative Sozialforschung, Band 2, Methoden und Techniken, S.202 ff, Weinheim: Belz, Psychologie Verlags Union.
12 Stokar von Neuforn, D., Thomaschewski, J.: Enhanging the motivation to learn through decoding text based communication reception in virtual learning environments – Case Study for the online course of media informatics, EDEN 2006 Annual Conference, E-Competences for Life, Employment and Innovation, Vienna University of Technology, Austria, 06.2006.

Atmosphere, Emotions (63.82%)	Address, dismissal, capital and small writing, length of the message, layout, vocabulary latency, colours and graphics, font, links and attachments, length of the sentence, sentence construction, spelling, questions, grammar, writing style, slang, abbreviations, extensions, filling words, expressions of strength, naming, metaphors, repetitions, smilies, punctuation marks, answering of questions, adjectives, questions about the condition, highlighting, compensated phonology, personal writing
Gender (46.80%)	capital and small writing, length of the message/sentence, colours and graphics, subject heading, sentence construction, writing style, formal writing, abbreviations, extensions, filling words, naming, metaphors, smilies, jokes, answering of questions, adjectives, questions about the condition, highlighting, reasoning of the message, personal writing, factual level
Interest (36.17%)	Address, length of the message, dismissal, layout, latency of the answers, links and attachments, answers in the email, senctence length, sentence construction, spelling, questions, answering of questions, factual level, questions about condition, jokes, reasoning of the message
Relational Aspect (31.91%)	Address, message length, dismissal, capital and small writing, layout, writing style, formal way of writing, slang, abbreviations, naming, smilies, factual level, jokes, formal/informal address, personal writing
Competence, education (25.53%)	Length of the message, layout, colours and graphics, links and attachments, sentence construction, spelling, grammar, writing style, slang, extensions, vocabulary
Time (23.40%)	Address, length of the message, layout, latency of the answers, font, links and attachments, answers in the email, sentence length, abbreviations, adjectives, answering of questions
Status, role, social Background (1.27%)	Address, signature, sentence construction, writing style, formal way of writing, slang, abbreviations, foreign words, punctuation marks
Appreciation (21.27%)	Address, dismissal, layout, latency of the answers, subject heading, sentence construction, questions, formal way of writing, extensions, apologies
Degree of trust/distance (12.76%)	Address, dismissal, capital and small writing, layout, grammar, personal writing
Taste/style /interests (6.38%)	Layout, font, slang
Identification of the person (6.38%)	Topic and reason of the message, font,

Figure 2: Illustration of the indicators for the evaluation of the text based conversation situation and their respective text based characteristics including the distribution by percentage among the text based characteristics.

Metagame Strategies of Nation-States, with Application to Cross-Strait Relations

Alex Chavez[1] and Jun Zhang[2,*]

[1]achavez@umich.edu, Department of Psychology, University of Michigan, USA
[2]junz@umich.edu, Department of Psychology, University of Michigan, USA

Abstract Metagames (Howard, 1968), a class of formal models in game theory, model the behavior of players who predict each other's conditional strategies recursively. We present a framework for three-player games based on metagame theory. This framework is well-suited to the analysis of nation-states, especially when the analyst wishes to make few assumptions about the level of recursive reasoning and the preference orderings of players.

1 Introduction

Much of game theoretic analysis relies on the concept of the Nash equilibrium, an outcome from which no player can deviate unilaterally and increase her or his payoff. As a solution concept, however, the Nash equilibrium is often inadequate for modeling stable outcomes in real-world applications. One well-known example is behavior in the Prisoner's Dilemma, pictured in Table 1. Although the unique Nash equilibrium in this game is for both players to defect, participants in experimental settings often cooperate.

Table 1 Prisoner's Dilemma. The number of the left in each cell (outcome) is Player 1's payoff, and the number on the right is Player 2's payoff.

		Player 2	
		Cooperate	Defect
Player 1	Cooperate	3,3	0,5
	Defect	5,0	1,1

A recent approach to modeling such discrepancies is to systematically modify the payoff function to include other-regarding preferences such as inequity aversion (Fehr and Schmidt, 1998) or conformity to social norms (Bicchieri, 2006). These approaches have enjoyed considerable success in modeling behavior in the laboratory setting. For political analysis, however, a more successful approach is to modify the players' strategies instead of their payoffs. Howard (1968)'s theory of metagames (see also Thomas, 1986) accomplishes this by assuming that one or more players use foresighted, conditional metagame strategies. Using the Prisoner's Dilemma as

---

* Currently on IPA assignment to AFOSR from the University of Michigan.

an example, Player 1's metagame strategies are of the form, "Cooperate if Player 2 cooperates; but defect if Player 2 defects." We refer to these as level-1 strategies. Furthermore, Player 2's strategy set builds off of Player 1's, and includes those such as "If Player 1 uses the above conditional strategy, cooperate; if player 1 uses a different conditional strategy, defect." We refer to these as level-2 strategies. By specifying the number of levels of nested strategies and the order in which players form their strategies, one generates the *metagame* corresponding to any normal-form *base game*.

Howard's proposed solution concept, the metaequilibrium, is an outcome in the base game that corresponds to a Nash equilibrium in the metagame. In the Prisoner's Dilemma, for example, because there are pairs of level-1 and level-2 metagame strategies that are Nash equilibria in the metagame and that correspond to (cooperate, cooperate) in the base game (Thomas, 1986), (cooperate, cooperate) is a metaequilibrium. An appealing property of metagames is that the inclusion conditional strategies higher than level-n (where n is the number of players) in the strategy set does not change the set of metaequilibria. Thus, there is no need to examine games with conditional strategies beyond level-n.

The metagame approach is useful especially when 1) the modeler is uncertain about the level of conditional strategies being employed by various players, or 2) the players are known to be using higher-level conditional strategies that are not captured in the base game. Both of these conditions often apply to situations of multinational conflict, and Taiwanese independence from mainland China is no exception. For this reason, combined with the fact that the status quo in the situation of Taiwanese independence is not a Nash equilibrium (as we will show in the following section), we use the metagame framework.

All metagames are constructed from a normal form base game. Therefore, we begin in Section 2.1 by formulating the situation of cross-strait relations as a metagame. To specify the game, we define the set of *players*, their *strategies*, and their *preferences over outcomes*. In Section 2.2, we present the formal solution concept of the Nash equilibrium, and show that the status quo is not contained in the set of Nash equilibria. Section 3 defines metagame strategies and the metaequilibrium. Then, Section 4 illustrates metagame analysis by analyzing in detail a particular preference combination. Finally, Section 5 concludes with a discussion of robustness of metaequilibria across many different possible preference orderings.

2 The Base Game

2.1 Outcomes and Preferences

We model the situation of cross-strait relations as the three-player normal-form game, Γ, in which: China (c), can wage war (*War*) or not (*NoWar*); Taiwan (t) can declare independence (*Independence*) or not (*NoIndependence*); and the United

States (u) can support Taiwan (*Support*) or remain neutral (*NoSupport*). These strategies jointly define the set of $2 \times 2 \times 2 = 8$ *outcomes* (see Table 2).

Each player i's preferences over these outcomes are modeled by a *payoff function*, π_i, which assigns real numbers to the outcomes in a manner that preserves the ordering of preferences. Thus, if Taiwan prefers outcome B to A, then $\pi_t(B) \geq \pi_t(A)$.

Table 2 Cross-strait Relations as a Three-player Game.

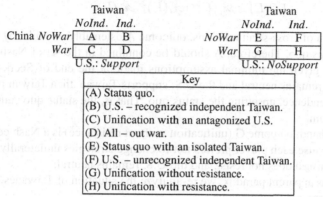

		Taiwan				Taiwan	
		NoInd.	Ind.			NoInd.	Ind.
China	NoWar	A	B		NoWar	E	F
	War	C	D		War	G	H
		U.S.: *Support*				U.S.: *NoSupport*	

Key
(A) Status quo.
(B) U.S. – recognized independent Taiwan.
(C) Unification with an antagonized U.S.
(D) All – out war.
(E) Status quo with an isolated Taiwan.
(F) U.S. – unrecognized independent Taiwan.
(G) Unification without resistance.
(H) Unification with resistance.

A typical assumption is that China prefers outcome G, unification without Taiwanese resistance or U.S. support, to all other outcomes. Taiwan, on the other hand, ranks the outcome of a U.S.–recognized independent state (B) the highest, but prefers capitulation (G) over a forceful invasion by China (H). Finally, when the stakes are low, the U.S. generally prefers to avoid conflict with China, and favors G over C.

Thus, we assume only the following partial preference orderings:

- $\pi_c(G) \geq \pi_c(E)$: When an unsupported Taiwan does not declare independence, China prefers unification (G) over staying neutral (E).
- $\pi_t(G) \geq \pi_t(H)$: Taiwan prefers to capitulate when China declares war and the U.S. remains neutral.
- $\pi_t(B) > \pi_t(A)$: Taiwan strictly prefers to declare independence if China remains neutral and the U.S. supports Taiwan.
- $\pi_u(G) \geq \pi_u(C)$: The U.S. prefers to avoid conflict with China, given that Taiwan does not declare independence.

2.2 Nash Equilibrium and Stability in the Situation of Taiwanese Independence

An outcome is *rational* for a player if she cannot change strategies without decreasing her payoff, holding the other players' strategies fixed. An outcome that is rational for all players is a *Nash equilibrium*. Equivalently, an outcome from which no player can unilaterally deviate and increase her payoff is a Nash equilibrium. Formally, the set of rational outcomes for each player $i = c, t, u$ in Γ is

$$\mathscr{R}_i(\Gamma) \equiv \{s \mid \pi_i(s) \geq \pi_i(s') \; \forall s, s' \in S\}, \tag{1}$$

where s and s' are outcomes in S. The set of Nash equilibria of Γ is

$$EQ(\Gamma) \equiv \mathscr{R}_c(\Gamma) \cap \mathscr{R}_t(\Gamma) \cap \mathscr{R}_u(\Gamma). \tag{2}$$

In the situation of cross-strait relations, outcome A ostensibly is stable, by virtue of being the status quo. Therefore, A should be contained in the set of Nash equilibria. However, given the minimal assumptions made at the end of Section 2.1, whenever China remains neutral and the U.S. supports Taiwan, then Taiwan prefers to declare independence, upsetting the status quo. That is, the status quo cannot be a Nash equilibrium.

On the other hand, outcome G (unification without resistance) is a Nash equilibrium. This is because each player prefers not to change strategies unilaterally when China unifies, Taiwan does not resist, and the U.S. remains neutral.

To resolve this apparent paradox, we re-analyze the situation of Taiwanese independence as a three-player metagame.

3 A Metagame Analysis

3.1 Metagame Strategies

Each metagame, $k\Gamma$, is constructed from the base game Γ and is identified by its *title k*, a string of 1, 2, or 3 players, which we denote respectively as k_1, $k_2 k_1$, and $k_3 k_2 k_1$ (where $k_i \in \{c, t, u\}$ are the players).[2] We refer any players not in the title and their metagame strategies as being of level-0, whereas each player k_n and k_n's metagame strategies are of level-i, where $1 \leq i \leq 3$. In the game $c\Gamma$, for example, China is the level-1 player, and Taiwan and the United States are level-0 players. In the game $utc\Gamma$, China is the level-1 player, Taiwan the level-2 player, and the United States the level-3 player. The metagame Γ is a special case in which all players are of level-0.

Metagame strategies are:

- for level-0 players, the same as their base game strategies;
- for level-i players ($0 < i \leq 3$), conditional strategies that specify the base game strategy to play for each combination of:

---

[2] In fact, strings with repetitions of any finite length are allowable (e.g., ttcutuuc). However, Howard (1968) proved that such strings may be collapsed by deleting all but the rightmost appearance of each player (so that ttcutuuc becomes tuc) without affecting the set of metaequilibria.

- the base game strategies of any other level-0 players,
- the metagame strategies of any other level-r players, where $0 < r < i$, and
- the base game strategies of any other level-l players, where $i < l$.

For example, consider China's strategies in the metagame $c\Gamma$ (in which China is of level-1, and Taiwan and the United States are of level-0). The strategies of China must specify whether to go to war, for each combination of the level-0 players' strategies. An example of one of China's metagame strategies is "*War* if (*NoIndependence,NoSupport*), but *NoWar* otherwise." Table 3 depicts this metagame strategy.

Table 3 An Example of China's Metagame Strategies in $c\Gamma$.

If		Then
(Taiwan)	(U.S.)	(China)
NoIndependence	*Support*	*NoWar*
NoIndependence	*NoSupport*	*War*
Independence	*Support*	*NoWar*
Independence	*NoSupport*	*NoWar*

Consider another metagame, $uc\Gamma$, in which the U.S. is of level-2, China of level-1, and Taiwan is of level-0. By definition, the level-1 player's metagame strategies specify which base game strategy to play for each combination of the base game strategies of the level-0 and level-2 players; thus, China has the same strategies as in 1Γ. The level-2 player's strategies, however, specify which base game strategy to play for each combination of the level-0 player's base game strategies, and the level-1 player's metagame strategies. The U.S. has $2^{32} \approx 4.3$ billion metagame strategies. An example of one such strategy is, "Support Taiwan only if Taiwan declares independence, and China uses a metagame strategy that specifies not to go to war if Taiwan declares independence."

In $tuc\Gamma$, the metagame strategies of the U.S. and China are the same as in $uc\Gamma$, except that Taiwan's are replaced by ones that specify a strategy for each combination of the U.S.'s and China's metagame strategies. Because there are $2^4 \times 2^5 = 512$ such combinations, the level-3 player (Taiwan) has 2^{512} metagame strategies.

3.2 Solution Concepts in the Metagame

3.2.1 Metagame Strategy Resolution and the Metaequilibrium

The metagame strategy of the highest level player can be resolved immediately as a base game strategy. Subsequently, the next highest level player's metagame strategy can also be resolved as a base game strategy, and so on, until all metagame strategies have been resolved, resulting in the identification of the outcome in the base game corresponding to the outcomein the metagame.

Having defined strategies and outcomes in the metagame, we now turn to the concepts of metarational outcomes and metaequilibrium. A base game outcome s is metarational for player i in $k\Gamma$, if there exists at least one metagame outcome that 1) resolves as s, and 2) is rational for i in $k\Gamma$. A base game outcome s is a metaequilibrium of $k\Gamma$ if it is metarational for all players. Also, it can be proven that if s is a metaequilibrium of $k\Gamma$, then that there exists at least one metagame outcome which 1) resolve as s and 2) is a Nash equilibrium in $k\Gamma$.

3.2.2 Identifying Metaequilibria

The number of outcomes in metagames with three players in the title is astronomical: $2^4 \times 2^{32} \times 2^{512}$. Although the Nash equilibria metagames with only player in the title can be found by a brute force search, there are simply too many outcomes to search over when two or more players are in the title. Fortunately, Howard (1971) proved the following theorem, which allows us to identify metaequilibria in the base game without identifying the corresponding Nash equilibria in the metagame.

Theorem 1 (Metarationality Theorem (Howard, 1971)). *For the metagame $k\Gamma$, let L_i equal the set of players to the left of i in the title if i is in the title, or equal the set of all players in the title if i is not in the title. Let R_i equal the set of players to the right of i in the title if i is in the title, or equal the empty set $\emptyset$ if i is not in the title. Let U_i be the set of players not in L_i, R_i, or $\{i\}$. Let S_i be the strategy set of i, and let S_{L_i}, S_{R_i}, and S_{U_i} be the respective joint strategy sets of L_i, R_i, and U_i. Finally, note the base game outcome s^* can be rewritten as $(s_{L_i}^*, s_i^*, s_{R_i}^*, s_U^*) \in S_{L_i} \times S_i \times S_{R_i} \times S_{U_i}$. Then, s^* is metarational in $k\Gamma$ for player i if*

$$\pi_i(s^*) \geq \min_{s_{L_i} \in S_{L_i}} \max_{s_i \in S_i} \min_{s_{R_i} \in S_{R_i}} \pi_i(s_{L_i}, s_i, s_{R_i}, s_{U_i}^*). \tag{3}$$

3.2.3 Symmetric Metaequilibrium

In addition, Howard (1968) proved that if an outcome is metarational in Γ, then it is metarational in $k_1\Gamma$. Furthermore, if an outcome is metarational in $k_1\Gamma$, then it is also metarational in $k_2 k_1 \Gamma$, and so on, so that metarational outcomes are nested in higher-level games. Since metaequilibria are simply outcomes which are metarational for all players, this implies that the metaequilibria are also nested in higher-level games. For example, the set of metaequilibria of 321Γ contains those in 21Γ, which in turn contains those in 1Γ, which in turn contains those in Γ.

Thus, to find all metaequilibria, we need only look at all games with titles of length three. This idea motivates an additional solution concept, the set of symmetric metaequilibria. A symmetric metaequilibrium is a base game outcome that corresponds to at least one metaequilibrium in $k\Gamma$, for *all* possible titles k. Such an equilibrium is robust in that it is stable for any arrangement of the players' conditional strategy levels.

In the following section, we use (3) to identify symmetric metaequilibria. But because the Metarationality Theorem only identifies metaequilibria, and not their corresponding outcomes in the metagame, we use brute force to find the latter in the metagames with level-1 players. Knowing which higher-level strategy combinations correspond to metaequilibria will, we hope, help elucidate the type of reasoning involved in the metagame.

4 Metaequilibria in the Taiwanese independence Game

In this section, we analyze the metagame corresponding to the preference ordering combination (GCHEABFD, BFADECGH, FBAEDHGC). Table 4 lists the metaequilibria for each possible title, whereas Tables 5–8 show the supporting metagame outcomes for Γ, $c\Gamma$, $t\Gamma$, and $u\Gamma$.

Table 4 Metaequilibria by Title for the Preference Ordering Combination: GCHEABFD, BFADECGH, FBAEDHGC.

Metagame	Metaequilibria
Γ	G
1Γ	BG
2Γ	BG
3Γ	FG
13Γ	BFG
23Γ	BFG
32Γ	BFG
12Γ	ABG
21Γ	ABG
31Γ	BG
132Γ	ABEFG
123Γ	ABEFG
212Γ	ABEFG
231Γ	ABEG
312Γ	ABEG
321Γ	ABEG
Symmetric:	ABEG

Table 5 Equilibria in the Base Game.

Strategies			Equilibrium Outcome
China	Taiwan	U.S.	
War	NoIndep.	NoSupport	$\rightarrow$ G

235

Table 6 Nash Equilibria in $c\Gamma$ and Corresponding Metaequilibria in the Base Game.

Equilibrium Strategies in the Metagame			Metaequilibrium
China	Taiwan	U.S.	
War if (*NoIndep.*, *Support*) War if (*NoIndep.*, *NoSupport*) Either if (*Indep.*, *Support*) War if (*Indep.*, *Support*)	*NoIndep.*	*NoSupport*	→ G
Either if (*NoIndep.*, *Support*) Either if (*NoIndep.*, *NoSupport*) NoWar if (*Indep.*, *Support*) War if (*Indep.*, *Support*)	*NoSupport*	*NoIndep.*	→ G

Table 7 Nash Equilibria in $t\Gamma$ and Corresponding Metaequilibria in the Base Game.

Equilibrium Strategies in the Metagame			Metaequilibrium
China	Taiwan	U.S.	
War	*Either* if (*NoWar*, *Support*) *NoIndep.* if (*NoWar*, *NoSupport*) *Either* if (*War*, *Support*) *NoIndep.* if (*War*, *NoSupport*)	*NoSupport*	→ G
NoWar	*Indep.* if (*NoWar*, *Support*) *Indep.* if (*NoWar*, *NoSupport*) *NoIndep.* if (*War*, *Support*) *Either* if (*War*, *NoSupport*)	*Support*	→ B

Table 8 Nash Equilibria in $u\Gamma$ and Corresponding Metaequilibria in the Base Game.

Equilibrium Strategies in the Metagame			Metaequilibrium
China	Taiwan	U.S.	
War	*NoIndep.*	*Either* if (*NoWar*, *NoIndep.*) *Either* if (*NoWar*, *Indep.*) *NoSupport* if (*War*, *NoIndep.*) *NoSupport* if (*War*, *Indep.*)	→ G
NoWar	*Indep.*	*Either* if (*NoWar*, *NoIndep.*) *NoSupport* if (*NoWar*, *Indep.*) *Either* if (*War*, *NoIndep.*) *Support* if (*War*, *Indep.*)	→ F

For example, consider $c\Gamma$ (see Table 6), in which China is the level-1 player. The outcome in which Taiwan does not declare independence, the United States does not support Taiwan, and China invades (G) is stable as long as China's metagame strategy *threatens* to invade if Taiwan or the U.S. unilaterally changes their strategy. This is because if Taiwan switches strategies and declares independence, outcome H would result, which Taiwan does not prefer to G. Similarly, if the U.S. switches strategies and supports Taiwan, outcome C would result, which the U.S. does to prefer to G. Of course, because G is China's preferred outcome, China would not change strategies.

On the other hand, the outcome of U.S.-supported Taiwanese independence (B) is stable if China's metagame strategy threatens go to war in the case that the U.S. does not support Taiwan. Although this might appear to be a strange equilibrium, it is stable because 1) China cannot improve by going to war, once the Taiwan and the U.S. have jointly used the above strategies, and 2) if China were not to threaten war, then the U.S. would not support Taiwan, and the outcome would not be an equilibrium.

In $u\Gamma$ (see Table 8), G is stable as long as the U.S. threatens not to support Taiwan if Taiwan declares independence. Similarly, Taiwanese independence not supported by the U.S. (F) is stable as long as the U.S. threatens to support Taiwan in the case that China goes to war.

Finally, in the metagame $tu\Gamma$, for example, the strategies of China and the U.S. would be of the same form as in Table 8. However, Taiwan's strategies would be of the form "If the U.S. uses the first strategy listed in Table 8 and China chooses *NoWar*, then declare independence; ...," with an if-then statement for every possible combination of the U.S.'s level-1 strategies and China's level-0 strategies.

5 Discussion and Conclusion

The status quo arises as a metaequilibrium in two level-2 metagames, and in all level-3 metagames (see Table 4), lending support to our choice of a metagame analysis of the situation of Taiwanese independence.

How robust is the set of metaequilibria we have identified? One approach to answering this question is to consider the general stability of our equilibrium predictions when the assumptions about the preference orderings are changed. To quantify this notion, we define the *signature* of a given set of preference orderings as the 15×8 binary matrix whose columns correspond to the eight outcomes of the game (A-H), and rows to the 15 possible titles (Γ, $t\Gamma$, $u\Gamma$, $c\Gamma$, $tu\Gamma$, $ut\Gamma$, $ct\Gamma$, $tc\Gamma$, $cu\Gamma$, $uc\Gamma$, $tuc\Gamma$, $tcu\Gamma$, $utc\Gamma$, $uct\Gamma$, $cut\Gamma$, $ctu\Gamma$). If an outcome is a metaequilibrium for a given title, the corresponding element in the matrix is defined to be 1; otherwise, the element is defined to be 0.

If the equilibrium prediction we have identified is robust, then changes in the preference orderings should not affect the corresponding signature. Thus, by pre-specifying a set of reasonable preference orderings (e.g., based on expert opinion), the relative frequencies of the resultant signatures index their robustness.

In conclusion, we have presented an application of a useful framework for analyzing three-player strategic situations. This framework is suited to the analysis of conflict and cooperation between nation-states, especially when the analyst wishes to make minimal assumptions about the level of reasoning and the preference orderings of players.

References

1. Bicchieri, C. (2006). *The Grammar of Society: the Emergence and Dynamics of Social Norms.* Cambridge University Press, Cambridge, Massachusetts.
2. Fehr, E. and Schmidt, K. (1999). A Theory of Fairness, Competition and Cooperation. *Quarterly Journal of Economics, 114*, 817-868.
3. Howard, N. (1968). *Theory of Metagames*, Ph.D. Thesis , University of London.
4. Howard, N. (1971). *Paradoxes of Rationality*. M.I.T. Press, Cambridge, Massachusetts.
5. Thomas, L. C. (1984). *Games, Theory and Applications*. Dover Publications, Inc., Mineola, New York.

Automating Frame Analysis

Antonio Sanfilippo, Lyndsey Franklin, Stephen Tratz, Gary Danielson, Nicholas Mileson, Roderick Riensche, Liam McGrath

{antonio.sanfilippo, lyndsey.franklin, stephen.tratz, gary.danielson, nicholas.mile- son, roderick.riensche, liam.mcgrath}@pnl.gov
Pacific Northwest National Laboratory, 902 Battelle Blvd., Richland, WA

Abstract Frame Analysis has come to play an increasingly stronger role in the study of social movements in Sociology and Political Science. While significant steps have been made in providing a theory of frames and framing, a systematic characterization of the frame concept is still largely lacking and there are no recognized criteria and methods that can be used to identify and marshal frame evidence reliably and in a time and cost effective manner. Consequently, current Frame Analysis work is still too reliant on manual annotation and subjective interpretation. The goal of this paper is to present an approach to the representation, acquisition and analysis of frame evidence which leverages Content Analysis, Information Extraction and Semantic Search methods to automate frame annotation and provide a systematic treatment of Frame Analysis.

1 Introduction

The objective of Frame Analysis is to understand the communicative and mental processes that explain how groups, individuals and the media try to influence their target audiences, and how target audiences respond. Since Goffman's initial exploration [9], frame analysis has become an important analytical component across the social and behavioral sciences [5].

The study of social movements in Sociology and Political Science is undoubtedly where Frame Analysis has had its strongest impact. The introduction of frame analysis in social movement theories was facilitated by the renewed interest in the social psychology of collective action during the early 1980s [6]. A series of seminal papers which rebuilt Goffman's initial insights within the study of collective action and with a more solid theoretical basis [7, 19] provided the initial impetus. In the next two decades, Frame Analysis grew very quickly to become a main component in Social Movement Theory [2, 11, 13, 20].

239

While significant steps have been made in providing a theory of frames and framing by Snow, Benford, Gamson, Entman and others, there are still no recognized criteria and methods that can be used to identify and marshal frame evidence reliably and in a time/cost effective manner [5]. Existing approaches to automated frame extraction [12, 14] are based on Entman's suggestion that frames can be "detected by probing for particular words" [3]. While producing interesting results, these approaches suffer from the lack of a more systematic approach to frame identification [16]. The goal of this paper is present an approach which addresses current limitations in the representation, automatic acquisition and analysis of frame evidence.

The main body of the paper is organized in 4 sections. Section 2 provides a review of the most important notions that have emerged in the relevant literature on frames and framing. In section 3, we show how these notions can be captured within a systematic automated approach to frame annotation, combining insights from linguistic theory with Content Analysis methods and Information Extraction capabilities. In section 4, we explore an application of the automated Frame Analysis approach presented, focusing on how a combined quantitative and qualitative analysis of frames can be made to support predictive intelligence analysis. We conclude with section 5, by discussing future developments of the approach presented.

2 Frame Analysis in Social Movement Theory

As practiced in theories of social movements, Frame Analysis has two main components: collective action frames and frame resonance. Collective action frames can be viewed as grammatical constructs which offer a strategic interpretation of issues with the intention of mobilizing people to act, e.g. *"Islam is the solution" said an Islamist politician*. Frame resonance describes the relationship between a collective action frame, the aggrieved community that is the target of the mobilizing effort, and the broader culture. In this paper we focus on collective action frames.

Two main perspectives have emerged on collective action frames. Gamson [6, 7] focuses on social-psychological processes that explain how individuals become participants in collective action. Snow and Benford [2, 19] concentrate on the relationship between social movement entrepreneurs and their potential constituencies. Table 1 provides a synopsis of these two perspectives. Entman [4] provides yet another characterization of collective action frames, as shown in Table 2. These three theories of collective action frames largely complement one another, presenting large areas of overlap. It would therefore be desirable that a coding schema for frame annotation be able to encompass the theoretical insights of all these approaches.

Table 1. Main perspectives on collective action frames.

Gamson	Snow and Benford
• *Identity*: ascertain aggrieved group with reference to shared interests and values • *Agency*: recognize that grieving conditions can be changed through activism • *Injustice*: identify individuals or institution to blame for grievances	• *Diagnostic frame*: tell new recruits what is wrong and why • *Prognostic frame*: present a solution to the problem suggested in the diagnosis • *Motivational frame*: give people a reason to join collective action

Table 2. Entman's characterization of collective action frames.

Substantive frame functions	Substantive frame foci
• Define effects or conditions as problematic • Identify causes • Convey moral judgment • Endorse remedies or improvements	• Political events • Issues • Actors

3 Automating Frame Annotation

Following Sanfilippo et al. [16], we analyze frames as speech acts [1, 18] that:
- convey a communicative *intent*;
- identify a frame *promoter*;
- may identify a frame *target*, and
- specify one or more *issues*.

This characterization equally leverages the various ways in which collective action frames have been treated in the literature.

- The notion of a frame *promoter* is used by Benford and Snow, corresponds to the result of Gamson's *identity* frame function, and overlaps with Entman's notion of *actor*.
- Communicative *intent* is implicit in the framing task classification provided by Gamson (*injustice, identity, agency*), Benford and Snow (*diagnostic, prognostic, motivational*), and Entman (*substantive frame functions*).
- Frame *target* partially overlaps with the result of Gamson's *injustice* frame function. The category *issues* is used by Entman.

- The frame categories *intent* and *issues* are divided into subcategories to provide a clearer mapping to the existing frame notions and facilitate identification, as shown in Table 3.

Table 3. Frame subcategories for intent and issues.

Communicative Intent	Issues
accept, accuse, assert, believe, correct, criticize, emphasize, explain, impute, judge, negate, reject, request, support, urge	economy, politics, social, law, military, administration, environment, security, religion

For each frame category, a specific set of guidelines is identified to reduce the level of subjective uncertainty in the assignment of frame annotations to text segments. For example, the identification of the *intent* subcategory "urge" in a text is established through occurrence of specific word meanings from WordNet (wordnet.princeton.edu), e.g. *urge#1, urge on#2, press#2, exhort#2*.

Sanfilippo et al. [16] demonstrate the viability of the resulting annotation scheme through inter-annotator agreement. A set of four human annotators expressing judgments about some 1800 frame ratings achieved an average Cohen kappa score of 0.70 ($z = 28$) and a Fleiss kappa score of 0.50 ($z = 46$). These are strong results given the number of categories (29) and annotators (4), and the level of category complexity.

Sanfilippo et al. [16] also show how an Information Extraction process can be crafted to identify frame categories in naturally occurring text automatically. This process combines a variety of language analysis capabilities including grammatical sentence analysis, named entity recognition, word sense and word domain disambiguation. The accuracy of automated frame was measured with reference to the 1800 human ratings used for computing inter-annotator agreement, using both the kappa score and precision/recall. The average Cohen kappa score between the automated annotator and the four human annotators was 0.52 ($z = 21$), the Fleiss kappa was 0.42 ($z = 50$), and the average precision/recall (F1) for frame detection was 0.74. Given the complexity of the task and the results of inter-annotator agreement, these results confirm that the automated frame annotation is viable.

We have recently augmented this process with two additional language analysis components to address coreference and temporal resolution. The coreference component is currently dealing with pronominal resolution, a vexing problem in the identification of frame components, e.g. an estimated 30% of frame promoters occur as pronouns in newswire type documents. We intend to extend the treatment of coreference to capture other types of nominal coreference (e.g. name aliasing and definite descriptions). The temporal resolution component provides a specification of both the time at which the frame event is reported and the time at which it occurs. Below is a sample output of the frame extraction process described.

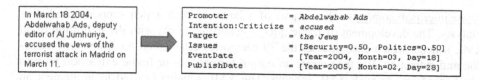

```
In March 18 2004,              Promoter              = Abdelwahab Ads
Abdelwahab Ads, deputy         Intention:Criticize  = accused
editor of Al Jumhuriya,        Target               = the Jews
accused the Jews of the        Issues               = [Security=0.50, Politics=0.50]
terrorist attack in Madrid on  EventDate            = [Year=2004, Month=03, Day=18]
March 11.                      PublishDate          = [Year=2005, Month=02, Day=28]
```

4 Marshaling Frame Content for Predictive Modeling

In this section, we will show how the Frame Analysis approach described in this paper can be used to support intelligence analysis. For ease of exposition, we will focus on the following task as use case:[1]

At the year-X elections in country-X, an underground radical organization named Group-X tripled the number of seats previously occupied to become the largest opposition bloc in the parliament. In the light of events leading to and following the year-X elections, establish whether the propensity for Group-X to adopt more secular views and become a civil political party is lower or greater than the propensity that the recent electoral success may lead to increased radicalization with consequent emergence of political violence.

Using insights from Social Movement Theory suggesting that political exclusion and unselective repression lead to increased radicalization [10], we hypothesize that strong evidence of contentious rhetoric about unjust repression and political exclusion would provide support for the possible emergence of political violence. We therefore set out to examine frames by group-X for signs of such a contentious rhetoric in the relevant period. First, we harvest a document corpus from the official Group-X web site for year-X though year-X+1 and obtain 619 documents. Then, we use the automated frame extraction process described in the previous section to extract frame content from the harvested data set. We analyze the frame evidence extracted through queries that utilize the frame categories established in Sanfilippo et al. [16] as search parameters. Our objective is to learn how predominant contentious frames of protest by Group-X are in the 619 documents harvested to assess radicalization potential.

In order to query the results of frame annotation, we have developed a semantically-driven and visually interactive search environment which combines the Alfresco content management platform (www.alfresco.com) with the SQL Server database

---

[1] US Government sponsored the work described in this paper. Therefore, certain data had to be omitted due to the terms of the sponsorship. These omissions do not affect the technical content and expository clarity of the paper.

(www.microsoft.com/sql). This is part of a larger system which supports predictive analysis. The development of frame query capabilities starts with the definition of an XML schema which encompasses the 29 categories established for frame representation and extraction. We then create an output format for our frame extraction process which conforms to such XML schema. The XML schema is used to implement an SQL Server database which makes it possible to load and query the XML output of the frame extraction process. The XML output includes pointers to the source documents from which the frame evidence is extracted. The source documents are indexed in Alfresco and can be retrieved using Alfresco web services by recovering document pointers through SQL queries. Semantic search is implemented in the form of SQL queries that enable parametric search and navigation over all the major frame category types, i.e. *promoter*, *intent*, *target*, *issues*, and *time*. The SQL queries are implemented in .NET and connect to the database via ODBC to achieve database independence, i.e. so that the queries would still work in the event the data were to be loaded on a different database (e.g. ORACLE). Finally, a web interface is implemented (in ASP.NET) to allow the user to issue queries and display results both in textual and graphical form, as shown in Figure 1 (frames are referred as "events of communication" and *intent* subcategories are listed in the "Actions" query box) .

Figure 1: Web interface to frame content search.

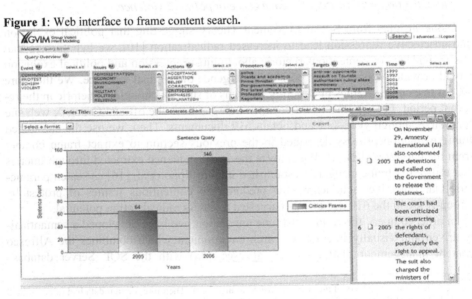

Using the search environment shown in Figure 1, we executed three queries to find all frames, "contentious" frames, and "negotiation" frames by Group-X. Contentious frames were extracted by selecting as search parameters the communicative *intent* subcategories: *accuse, correct, criticize, judge, negate, reject,* and *request*. Negotiation

244

frames were extracted by selecting as search parameters the communicative *intent* subcategories: *accept, explain,* and *support.* The results of these searches, shown in Figure 2, indicate that contentious and negotiation frames represent a modest proportion of all Group-X frames with an overall predominance of contentious frames, as illustrated in Table 4.

The overall predominance of contentious frames, both in the period preceding and following the elections, would suggest that there is an overall tendency for Group-X to prefer contentious to negotiation communicative strategies. The increase in the occurrence of contentious frames and decrease in the occurrence of negotiation frames moving from year-X, the year in which the parliamentary elections took place, to the following year suggests that electoral success is leading to increased radicalization. These conclusions are corroborated by the z-scores in Table 4, which indicate a notable increase from strong ($z = 2.28$) to very strong ($z = 6.16$) statistical significance for the increased distribution of contentious frames from year-X to year-X+1 (see explanation below). This information can be used within a predictive model of violent intent for Group-X to counter the hypothesis that winning the elections in year-X will reduce the rate of radicalization and consequently may not quell the propensity of Group-X to indulge in violent behavior.

Figure 2: Frame query results for Group-X in years X and X+1.

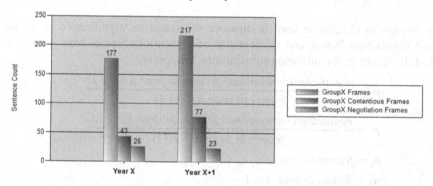

Table 4. Frame distribution for Group-X in years X and X+1 (z-scores compare contentious vs. negotiation frame proportions by year).

	Contentious Frames	**Negotiation Frames**	**z-score**
Year X	24% (43/177)	14% (26/177)	2.28
Year X+1	35% (77/217)	10% (23/217)	6.16

The z-scores in Table 4 express the difference between the proportions of contentious and negotiation frames as standard deviation values. These z-scores are computed as a test for comparing two binomial proportions [15: pp. 291-297], using the formula in (1) where

$$\hat{p}_1 = \frac{\textit{contentious frames in year X (X + 1)}}{\textit{frames in year X (X + 1)}}$$

$$\hat{p}_2 = \frac{\textit{negotiation frames in year X (X + 1)}}{\textit{frames in year X (X + 1)}}$$

$$p = \frac{\textit{contentious and negotiation frames in year X (X + 1)}}{\textit{frames in year X (X + 1)} \times 2}$$

$$q = 1 - p$$

$$n_1 = n_2 = \textit{frames in year X (X + 1)}.$$

Approximately, a z-score of 1.96 indicates a 95% confidence interval, and a z-score of 2.58 indicates a 99% confidence interval [15: pp. 292, 296].

$$(1) \quad z = \frac{\hat{p}_1 - \hat{p}_2}{\sqrt{pq\left(\frac{1}{n_1} + \frac{1}{n_2}\right)}}$$

The formula in (1) can be used to measure the statistical significance in the decrease of negotiation frames and the increase of contentious frames from year X to year X+1. In this case, the following substitutions are applied

$$\hat{p}_1 = \frac{\textit{negotiation (contentious) frames in year X (X + 1)}}{\textit{frames in year X (X + 1)}}$$

$$\hat{p}_2 = \frac{\textit{negotiation (contentious) frames in year X + 1 (X)}}{\textit{frames in year X + 1 (X)}}$$

$$n_1 = \textit{frames in year}$$

$$n_2 = \textit{frames in year X + 1.}$$

Table 5. Frame distribution for Group-X in years X and X+1 (z-scores compare contentious or negotiation frame proportions by year).

	Contentious Frames	Negotiation Frames
Year X	24% (43/177)	14% (26/177)
Year X+1	35% (77/217)	10% (23/217)
z-score	2.40	1.22

The z-scores in Table 5 strongly confirm the tendency for contentious frames to increase from year X to year X+1, but only weakly support the tendency for negotiation frames to decrease from year X to year X+1.

Finer grained assessments about the distribution of contentious and negotiation frames to support the likelihood estimation of increased radicalization can be made by restricting the range of *issues* selected in frame queries. For example, we can more specifically assess rhetoric about unjust repression by retrieving contentious frames which include issues such as *security* and *politics* rather than *environment* and *economy*. Finally, we wish to stress that the analysis presented here is a partial assessment of increased radicalization. A more comprehensive treatment should take into consideration additional factors such as indicators of political opportunities and constraints, resources and mobilization, leadership traits, and group entitativity.

5 Conclusions

We have presented an approach to the extraction and marshaling of frame signatures which enables the analysis of messaging strategies from document collections in a time and cost effective fashion. This work represents a significant evolution of frame extraction work, and can be profitably used to support predictive modeling tasks.

Moving forward, one major area of development that we are currently targeting regards the extraction and analysis of frames from documents which record direct speech, e.g. interviews, sermons and speeches. This extension will enable the recognition and analysis of frames even when the communicative *intent* verb is not overtly expressed. We expect that the automation of frame analysis in direct speech will require novel research and development in two main areas: the automatic identification of documents which record direct speech in text, and the creation and use of domain specific frame ontologies. Sanfilippo and Nibbs [17] provide some preliminary details about the development and use of domain specific frame ontologies to support the analysis of the communicative intent expressed by the speaker in direct speech.

Acknowledgements We would like to thank Paul Whitney and Christian Posse for advice on the the z-scores calculations described in section 4, and Elsa Augustenborg for constructive feedback on the web interface to frame content search. Many thanks also to Annie Boek, Linda Connell, Rebecca Goolsby, Barry Markovsky, Mansoor Moaddel, Jennifer O'Connor, Allison Smith, Paul Sticha and Steve Unwin for generous comments on previous oral and written versions of this paper.

References

1. Austin, J. (1962) How To Do Things with Words. Oxford University Press, Oxford, UK.
2. Benford, D. and R. Snow (2000) Framing Processes and Social Movements: An Overview and Assessment. Annual Review of Sociology, Vol. 26, pp. 611-639.
3. Entman, R. (1993) Framing: Toward Clarification of a fractured paradigm. Journal of Communication 43(4): 51-58.
4. Entman, R. (2004) Projections of power: Framing news, public opinion, and U.S. foreign policy. University of Chicago Press, Chicago, IL.
5. Fisher, K. (1997) Locating Frames in the Discursive Universe. Sociological Research Online, vol. 2, no. 3, http://www.socresonline.org.uk/ socresonline/2/3/4.html.
6. Gamson, W., B. Fireman and S. Rytina (1982) Encounters with Unjust Authority. Dorsey Press, Homewood, IL.
7. Gamson, W. (1988) Political Discourse and Collective Action. In Klandermans, B., H. Kriesi, and S. Tarrow (eds) International Social Movement Research, Vol. 1. JAI Press, London.
8. Gamson, W. (1992) Talking Politics. Boston: Cambridge University Press, New York, NY.
9. Goffman, E. (1974) Frame Analysis: An Essay on the Organization of Experience. Harper and Row, London, UK.
10. Hafez, M. (2003) Why Muslims Rebel: Repression and Resistance in the Islamic World. Lynne Rienner Publishers, Boulder, CO.
11. Noakes, J. and H. Johnston (2005) Frames of Protest: A Road Map to a Perspective. In Johnston, H., and J. Noakes (eds.) Frames of Protest: Social Movements and the Framing Perspective. Rowman & Littlefield, Lanham, MD.
12. Koenig, T. (2004). Reframing Frame Analysis: Systematizing the empirical identification of frames using qualitative data analysis software. ASA, San Francisco, CA, 8/14-17/04.
13. McAdam, D., S. Tarrow, C.Tilly (2001) Dynamics of Contention. Cambridge University Press, New York, NY.
14. Miller, M. (1997) Frame Mapping and Analysis of News Coverage of Contentious Issues. Social Science Computer Review, 15 (4): 367-78.
15. Ott, L., W. Mendenhall, R. Larson (1978) Statistics: A Tool for the Social Sciences. 2nd edition. Duxbury Press, North Scituate, MA.
16. Sanfilippo, A., A.J. Cowell, S. Tratz, A. Boek, A.K. Cowell, C Posse, and L. Pouchard (2007) Content Analysis for Proactive Intelligence: Marshaling Frame Evidence. In Twenty-Second AAAI Conference on Artificial Intelligence in Vancouver, BC, Canada.
17. Sanfilippo, A., and F. Nibbs (2007) Violent Intent Modeling: Incorporating Cultural Knowledge into the Analytical Process. Lab report No. 16806, PNNL, Richland, WA.
18. Searle, J. (1969) Speech Acts. Cambridge University Press, Cambridge, UK.
19. Snow, D., B. Rochford, S. Worden and R. Benford (1986) Frame Alignment Processes, Micro-mobilization and Movement Participation. American Sociological Review, vol. 51, no. 4, pp. 464 - 81.
20. Wiktorowicz, Q. (2004) Introduction: Islamic Activism and Social Movement Theory. In Wiktorowicz, Q. (ed.) Islamic Activism: A Social Movement Theory approach. Indiana University Press, Bloomington, IN.

Using Topic Analysis to Compute Identity Group Attributes

Lashon B. Booker[†] and Gary W. Strong[*]

[†] booker@mitre.org, The MITRE Corporation, McLean, Virginia
[*] gstrong@jhu.edu, Human Language Technology Center of Excellence, Johns Hopkins University, Baltimore, MD

Abstract Preliminary experiments are described on modeling social group phenomena that appear to address limitations of social network analysis. Attributes that describe groups independently of any specific members are derived from publication data in a field of science. These attributes appear to explain observed phenomena of group effects on individual behavior without the need for the individual to have a network relationship to any member of the group. The implications of these results are discussed.

1 Introduction

The dynamics of many group phenomena, such as fads and "infectious ideas", appear to exhibit the characteristics of phase-change-behavior (i.e., sudden changes in system behavior). We hypothesize that phase change behavior occurs in part because of a change in the ideas and beliefs that bind individuals with social identity groups and bias individual behavior. We have been conducting research aimed at exploring this hypothesis. The objective of this research is to develop a modeling framework that will allow us to investigate group recruitment processes, inter-group dynamics, and the effects of the group upon individual members' behavior. This paper describes preliminary experiments suggesting that, in some cases, identity group attributes are best modeled directly at the group level rather than as derivative properties of individual member attributes.

In our view, the attributes of social identity groups play a central role in modeling group dynamics. Consequently, we believe that models must include explicit representations of both the individual group members as well as the social identity groups to which individuals may belong. This differs from the traditional modeling approaches in social network analysis where models only explicitly represent individuals and their relationships, leaving groups to be represented as clusters of individuals. Such models are straightforward to derive and to analyze, but are insufficient to model sudden changes in the behavior of a population in which group identity plays a distinct role beyond member-to-member relationships. Garfinkel [3] describes an important example of this possibility, showing how groups can not only be leaderless and organization-less, but member-less for periods of time as well. Such a possibility is accounted for by the potential for persistence of group identity on the Internet over spans of time when there may, in fact, be no members in the group at all. If, as we claim, sudden changes in behavior are due, in part, to changes at the level of identity group, or class, then we need to have models with representations of both agents and the abstract groups to which agents may belong. This allows groups to have attributes and associated information without needing to have specific members and allows members to associate with groups without having to affiliate with any other individual members of the group, at least initially.

The first step in building models of this kind is to identify group attributes that are independent of particular members, allowing modeling of phenomena whereby individuals are drawn to attributes of the group and not specifically to any of its members, if there are any. Models that allow for differential attraction to group-level attributes offer a way around the limitation of social networking techniques that depend upon actual individual-to-individual relationships.

Our initial investigation of these ideas has focused on the role of identity groups in the formation of collaborative networks associated with scientific publications. This paper describes how we characterize identity groups and identity group attributes in this domain. We also present experiments showing that the group attributes provide better predictions of the contents of documents published by group members than attributes derived from the individual documents.

2 Identity Groups and Scientific Collaboration

An important goal of studies examining the collaborative networks associated with scientific publications is to understand how collaborative teams of co-authors build a social network that eventually forms an "invisible college" in some field of study [6]. We are particularly interested in understanding and modeling how identity groups influence the way teams are assembled. What are the important social identity groups in this setting? Publication venues are visible manifestations of some of those key social identity groups. People tend to publish papers in conferences where there is some strong relationship between their interests and the topics of the conference. Consequently, conference participants tend to have overlapping interests that give them a meaningful sense of social identity. There are many attributes that might be useful to characterize these groups, ranging from the various social and academic relationships between individuals to the organizational attributes of professional societies and funding agencies. One readily available source of information about a group is the corpus of documents that have been collectively published by group members. Given this document collection as a starting point, topics and frequently used keywords are an obvious choice as identity group attributes in this domain. The peer review system is a mechanism that ensures published papers include enough of the topics and keywords considered acceptable to the group as a whole. Authors that do not conform to these norms and standards have difficulty getting their papers published, and have difficulty finding funding for their work.

2.1 Topic Analysis

If topics are considered to be the identity group attributes of interest, it is natural to turn to topic analysis as a way of identifying the attributes characterizing each group. Topic analysis techniques have proven to be an effective way to extract semantic content from a corpus of text. Generative probabilistic models of text corpora, such as Latent Dirichlet Allocation (LDA), use mixtures of probabilistic "topics" to represent the semantic structure underlying a document [5]. Each topic is a probability distribution over the words in the corpus. The gist or theme of a topic is reflected in selected words having a relatively high probability of occurrence when that topic is prevalent in a document. Each document is represented as a probability distribution over the topics associated with the corpus. The more prevalent a topic is in a document, the higher its relative probability in that document's representation. We use an LDA model of the scientific document collections in our studies of collaboration networks

An unsupervised Bayesian inference algorithm can be used to estimate the parameters of an LDA model; that is, to extract a set of topics from a document collection and estimate the topic distributions for the documents and the topic-word distributions defining each topic. As illustrated in Figure 1, the algorithm infers a set of latent variables (the topics) that factor the word document co-occurrence matrix. Probabilistic assignments of topics to word tokens are estimated by iterative sampling. See [4] for more details. Once the model parameters have been estimated, the topic distribution for a new document can be computed by running the inference algorithm with the topic-word distribution fixed (i.e., use the topic definitions that have already been learned).

2.2 Document Classification

If topics are indeed identity group attributes for publication venues viewed as social identity groups, then these attributes ought to distinguish one group from another somehow. Different groups should have distinguishable topic profiles and the topic profile for a document should predict its group. Document classification experiments can be used to verify that identity group influence on published papers is reflected in document topic profiles.

The document descriptions derived from an LDA model are ideally suited to serve as example instances in a document classification problem. One outcome of estimating the parameters of an LDA model is that documents are represented as a probability distribution over topics. Each topic distribution can be interpreted as a set of normalized real-valued feature weights with the topics as the features. Results in the literature suggest that these features induced by an LDA model are as effective for document classification as using individual words as features, but with a big advantage in dimensionality reduction [1].

Regardless of what features we use, we would expect that documents from different topic areas would be distinguishable in a document classification experiment. A more interesting question is whether or not one set of features provides more discriminatory information than another. This is where our research hypothesis about group-level attributes becomes relevant. We hypothesize that in many cases the group identity is most effectively represented by group-level attributes. What are group-level attributes associated with the document collections being considered here?

The standard approach to using LDA is to compute topics for the entire corpus that account for the word co-occurrences found in each individual document. Our modeling emphasis on identity groups led us to consider an alternative approach that focuses instead on word co-occurrences at the group level. A very simple procedure appears to make this possible. First we aggregate the documents affiliated with an identity group into a single mega-document. Then, we use LDA to compute topics for this collection of mega-documents. Since the resulting topics account for word co-occurrences in the mega-documents, they are attributes of the group and not attributes associated with the individual documents. However, these group-level topics can also be used as attributes to characterize each individual document, since all topics are represented as probability distributions over the same vocabulary of words. We hypothesize that the topics computed in this manner directly capture the attributes associated with each identity group as a whole and in some sense should be better than attributes derived from the word co-occurrences in individual documents.

3 Empirical Results

In order to test this hypothesis, we conducted a document classification experiment in which topics extracted from an unlabeled body of text were used to predict the social identity group (that is, publication venue) of that document. The document collection for this experiment was selected from a dataset containing 614 papers published between 1974 and 2004 by 1036 unique authors in the field of information visualization [2]. The dataset did not include the full text for any of the documents, so we worked with the 429 documents in the dataset that have abstracts. The title, abstract and keywords of each entry constituted the "bag of words" to be analyzed for each individual document.

Table 1. Publication venues and number of documents.

Source name	count
Proceedings of the IEEE Symposium on Information Visualization	152
IEEE Visualization	32
Lecture Notes in Computer Science (LNCS)	22
Conference on Human Factors in Computing Systems (SIGCHI)	21
ACM CSUR and Transactions (TOCS,TOID,TOG)	21
Symposium on User Interface Software and Technology (UIST)	17
IEEE Computer Graphics and Applications	13
IEEE Transactions	13
International Conference on Computer Graphics and Interactive Techniques (CGIT)	12
Communications of the ACM (CACM)	10
Advanced Visual Interfaces (AVI)	10
Other	106

Several publication venues were represented here. We arbitrarily decided to consider each venue having 10 or more publications in the dataset as an identity group. This requirement provided some assurance that enough information was available to determine useful topic profiles for each group. Papers that did not belong to a venue meeting this minimum requirement were lumped together into a default group called "Other". Table 1 lists the identity groups and the number of documents associated with them. Given that the field of information visualization is itself a specialized identity group, it is not obvious that the smaller groups we specify here will have any distinguishable properties. It is also not obvious that broadly inclusive groups like the IEEE Symposium on Information Visualization or the "Other" category will have any distinguishable properties since they include papers from all the relevant topic areas in the field.

The selected documents were preprocessed to convert all words to lower case and remove all punctuation, single characters, and two-character words. We also excluded words on a standard "stop" list of words used in computational linguistics (e.g., numbers, function words like "the" and "of", etc.) and words that appeared fewer than five times in the corpus. Words were not stemmed, except to remove any trailing "s" after a consonant. This preprocessing resulted in a vocabulary of 1405 unique words and a total of 31,256 word tokens in the selected documents.

Two sets of topic features were computed from this collection: one from the set of individual documents and one from the set of aggregated mega-documents associated with each group. The optimal number of topics that fits the data well without overfitting were determined using a Bayesian method for solving model selection problems [4]. The model with 100 topics had the highest likelihood for the collection of mega-documents and the 200 topic model was best for the collection of individual documents. The resulting feature vector descriptions of each document were then used as examples for training classifiers that discriminate one identity group from another. Examples were generated by running the LDA parameter estimation algorithm over the words in a document for 500 iterations and drawing a sample of the topic distribution at the end of the run. Ten examples were generated for each document[1], producing an overall total of 4290 examples for each feature set.

On each run of the experiment, a support vector machine classifier [7] was trained with a random subset of 75% of the examples, with the remaining 25% used for testing. For each group, we trained a "one-versus-the-rest" binary classifier. This means that the examples from one class became positive examples while the examples from all other classes were treated as negative examples in a binary classification task. The overall solution to the multi-class problem is given by following the recommendation of the binary classifier with the highest classification score on a given test example. The support vector machine used a radial basis function kernel along with a cost factor

---

[1] Since the algorithm and representation are probabilistic, different runs will produce different feature vectors. The "true" feature vector can be thought of as a prototype that is most representative of all possible sample vectors.

to compensate for the unbalanced number of positive and negative examples. The cost factor weighs classification errors so that the total cost of false positives is roughly equal to the total cost of false negatives For details about the cost factor, see [8].

Results averaged over 10 independent runs show that the group-level features produce a statistically significant improvement in overall classification accuracy over the document-level features on test data (88.7% accuracy versus 79.7% accuracy). Not surprisingly, the 2 most diverse classes ("IEEE Symposium on Information Visualization" and "Other") had the worst classification performance. The confusion matrix data shows that most classification errors were due to erroneous assignments to one of these classes.

Table 2 - Classification results

Identity Group	# Examples	Document-level Accuracy	Group-level Accuracy
ACM CSUR	210	88.91%	96.91%
AVI	100	74.30%	98.06%
CACM	100	80.11%	91.77%
CGIT	120	82.01%	100.00%
IEEE Comp Graphics	130	80.87%	94.40%
IEEE Symp on InfoVis	1520	77.49%	87.34%
IEEE Transactions	130	71.81%	87.60%
IEEE Visualization	320	80.03%	90.94%
LNCS	220	95.40%	97.12%
SIGCHI	210	87.35%	90.06%
UIST	170	93.68%	90.73%
Other	1060	75.31%	83.89%
Overall	4290	79.74%	88.70%

4 Summary and Conclusions

This paper shows that group-level attributes can provide better predictions of the contents of scientific documents published by group members than predictions based on attributes derived from the individual documents. It represents a modest step toward developing new approaches to modeling the effects of social identity groups on the behavior of individual members. The potential implications of this line of research could be far reaching. Social identity groups appear to adapt over time, store behaviors, and pass them on to new members. In this fashion, they may be important vehicles for cultural memory. Observed phenomena of sudden, mass effects of human behavior may be due, in part, to such effects of certain social groups upon individuals.

Social network analysis is currently a dominant paradigm for modeling social groups. Groups, as defined by social network analysis consist of individual actors tied together into networks defined by person-to-person relationships of one kind or another depending upon what's being modeled. Social network analysis is an alternative to more traditional social analyses that relate people to each other according to particular individual attributes, beliefs, or behaviors, characteristics that can are often derived from social surveys.

Data mining has proven to be a useful network discovery technique in social network analysis because of the ability to build social networks by pair-wise discovery of individual-to-individual relationships in different electronic databases. Such databases could be, for example, email messages or telephone calls between individuals, and strength of relationship could be defined as the number of communications found in the data. Data mining results enable the identification of social groups by noting the clusters of individuals who interact with each other based upon some pre-defined threshold of relationship strength. Such clusters are different from those based on surveys because they are representations of actual interacting groups, depending upon the coverage of the databases, rather than abstractions resulting from group parameter estimation based upon a population sampled in a survey.

Social networking approaches however are less useful when explaining group persistence over periods of time when there are no members or when explaining recruitment of members through means other than direct contact with existing members such as is done through advertising. This limitation of social network analysis is due to the dependency of social network clusters upon the data of interacting individuals. Social networks, defined using the normal techniques of social network analysis cannot easily represent groups that do not currently have members. While this may seem, on the face of it, to be unimportant, there is evidence as discussed in [3] for group effects on individual behaviors even during times when there are no group members to cause such behavior. Modern communication, including the Internet in particular, has promoted phenomena such as viral marketing, branding, infectious ideas, propaganda, political campaigns, and other similar types of phenomena. Common to these phenomena is the fact that they often seek to recruit individuals to a group as an abstract concept rather than recruit by developing new relationships with specific members of a group.

The approach we have taken in this paper identifies group attributes that are independent of particular members, allowing modeling of recruitment phenomena whereby individuals are drawn to attributes of the group and not specifically to any of its members, if there are any. Models, such as ours, that allow for differential attraction to group-level attributes offer a way around the limitation of social networking techniques that depend upon actual individual-to-individual relationships.

Models of group-level phenomena may also turn out to be helpful in explaining the development of new scientific fields, beyond the formation of teams of co-authors of scientific publications as studied by [6]. It is well known that a researcher's career will be heavily influenced by peer reviewers since reviews of papers submitted for publication as well as proposals for funding normally are drawn from an existing scientific community of experts. It has been observed that science proposals that lie outside of an existing community and its paradigms are often considered too risky to fund by National Science Foundation programs. The same conservatism with regard to straying from established disciplinary topics may also hold for many disciplinary scientific journals as well.

A group-related phenomenon of interest in science is, given what is perceived to be a natural conservatism related to peer review, how do interdisciplinary sciences get started? An extension of our hypothesis regarding social identity groups is that interdisciplinary groups form when topic drift of one or more fields occurs. That is, a given community of scientists may develop multiple subspecialties as represented by the topics about which different subsets of the group tend to publish. Such subsets may drift such that there is significant subgroup and associated topics such that, when papers or proposals are submitted, editors and program managers recognize the distinction and aim incoming manuscripts accordingly to the right subgroups such that appropriately-matched peer reviewers are chosen.

There is evidence that some new scientific areas, such as bioinformatics, have developed out of the intersection of the interests of different communities of researchers, such as molecular biology and computer science. Multi-group interactions may not always be as important as fractionation of an existing community into two or more separable communities over time, but it represents another way in which new topics and associated subgroups of experts may form. Research is yet to be done on whether or not group fractionation and recruitment is commonly found to be associated with the development of new scientific fields.

References

1. Blei D, Ng A, Jordan M (2003) Latent Dirichlet Allocation. *Journal of Machine Learning Research 3*:993-1022.
2. Börner K, Dall'asta L, Ke W, Vespignani A (2005) Studying the Emerging Global Brain: Analyzing and Visualizing the Impact of Co-Authorship Teams. *Complexity 10(4)*: 57-67.
3. Garfinkel S. (2003) Leaderless resistance today. *First Monday 8(3)*. http://www.firstmonday.org/issues/issue8_3/garfinkel/index.html.
4. Griffiths T, Steyvers M (2004) Finding scientific topics. *Proceedings of the National Academy of Science 101*: 5228-5235.
5. Griffiths T, Steyvers M, Tenenbaum, J (2007) Topics in semantic representation. *Psychological Review 114(2)*:211-244.
6. Guimera R, Uzzi B, Spiro J, Amaral, L (2005) Team Assembly Mechanisms Determine Collaboration Network Structure and Team Performance. *Science 308(5722)*: 697-702.

7. Joachims T (1999) Making large-scale SVM learning practical. In: Schölkopf B, Burges C, Smola A (Eds) Advances in Kernel Methods – Support Vector Learning. MIT Press.
8. Morik K, Brockhausen P, Joachims T (1999) Combining statistical learning with a knowledge-based approach – A case study in intensive care monitoring. In: Proceedings of the 16th International Conference on Machine Learning 268-277. Morgan Kaufmann, San Francisco.

Biographies

Huan Liu, ASU

Huan Liu is on the faculty of Computer Science and Engineering at Arizona State University. He received his Ph.D. and MS. (Computer Science) from University of Southern California, and his B.Eng. (Computer Science and Electrical Engineering) from Shanghai Jiaotong University. His research interests include machine learning, data and web mining, social computing, and artificial intelligence, investigating search and optimization problems that arise in many real-world applications with high-dimensional data such as text categorization, biomarker identification, group profiling, searching for influential bloggers, and information integration. He publishes extensively in his research areas, including books, book chapters, encyclopedia entries, as well as conference and journal papers.

John J. Salerno, AFRL

John Salerno has been employed at the Air Force Research Laboratory for the past 27 years where he has been involved in many research disciplines. He began his career in the Communications Division where he was involved in and ran a number of research efforts in both voice and data communications networks. In 1989 he moved from the Communications Division into the Intelligence and Exploitation Division. Since then he has been involved in research in distributed heterogeneous data access architectures, web interfaces, and in the further definition and implementation of fusion to various application areas. He has designed, built and fielded a number of operational systems for the Intelligence Community. He holds two US patents and has over 50 publications. John received his PhD from Binghamton University (Computer Science), a MSEE from Syracuse University, his BA (Mathematics and Physics) from SUNY at Potsdam and an AAS (Mathematics) from Mohawk Valley Community College. He is an AFRL Fellow.

Michael J. Young, AFRL

Michael J. Young is Chief of the Cognitive Systems Branch of the Air Force Research Laboratory where he directs a variety of research on human factors, adversary modeling, and cultural research. During his twenty-two year career his primary research goal has been improving the representations of human behavior in training systems, command post exercises, and decision aids. He has over 20 publications. He received a PhD from Miami University (Experimental Psychology), and a M.A. (Psychobiology) and an A.B. (Psychology) from the University of Michigan.

Author Index